Thomas Püttker

Effects of Habitat Fragmentation on Small Mammals of the Atlantic Forest, Brazil

Thomas Püttker

Effects of Habitat Fragmentation on Small Mammals of the Atlantic Forest, Brazil

Use of Vegetation Structures, Density, Movement Patterns and Parasite Load of Selected Species

VDM Verlag Dr. Müller

Imprint

Bibliographic information by the German National Library: The German National Library lists this publication at the German National Bibliography; detailed bibliographic information is available on the Internet at http://dnb.d-nb.de.

Cover image: www.purestockx.com

Publisher:
VDM Verlag Dr. Müller Aktiengesellschaft & Co. KG , Dudweiler Landstr. 125 a, 66123 Saarbrücken, Germany,
Phone +49 681 9100-698, Fax +49 681 9100-988,
Email: info@vdm-verlag.de

Zugl.: Hamburg, UHH, Diss., 2007

Produced in USA and UK by:
Lightning Source Inc., La Vergne, Tennessee, USA
Lightning Source UK Ltd., Milton Keynes, UK
BookSurge LLC, 5341 Dorchester Road, Suite 16, North Charleston, SC 29418, USA

ISBN: 978-3-639-01820-2

Table of contents

SUMMARY

Habitat fragmentation is a major threat to biodiversity all over the world and especially in tropical rainforests. The coastal Atlantic forest, Brazil (Mata Atlântica), is one of the 25 biodiversity hotspots of the world and at the same time one of the world's most threatened environments. The Atlantic forest is home to over 20,000 plants and over 1360 vertebrate species. Additionally, its high degree of endemic species (about 40 % of inhabiting plant and vertebrates) outlines its high value for biodiversity conservation. Only about 7 % of the original forest remains, which does not only include primary forest but consists in large parts of secondary forest remnants in different stages of regeneration. These secondary forests play an important role in species conservation when primary forests are limited.

Small mammal species respond in different ways to the fragmentation of the forest habitat. Habitat generalists are - in contrast to specialist species - not impaired by fragmentation of the forest habitat. The reasons for these differences between small mammals of the Atlantic forest were unknown and my investigations aimed to emphasize knowledge about the ecology of these species. I investigated the effects of fragmentation on movement patterns, population density, use of vegetation structures and on parasite load of selected small mammal species of the Atlantic forest in the vicinity of the city of São Paulo, Brazil. The underlying question in all investigations was: Are these parameters of small mammal ecology influenced by habitat fragmentation?

Focus species were chosen based on the results of previous research by our Brazilian co-operators (Rodents: *Akodon montensis* (Grass mouse), *Oligoryzomys nigripes* (Black-footed pygmy rice rat), *Delomys sublineatus* (Pallid Atlantic forest rat); Marsupials: *Marmosops incanus* (Gray slender mouse opossum), *Gracilinanus microtarsus* (Brazilian gracile mouse opossum). The rodent *D. sublineatus* and the two marsupial species are endemic species of the Atlantic forest. *A. montensis*, *O. nigripes*, and *G. microtarsus* seemed to be more generalistic species with respect to habitat fragmentation. On the other hand, *M. incanus* and *D. sublineatus* were more sensitive to fragmentation effects.

The study area consisted of a continuous forest area of approximately 10,000 ha and a neighbouring fragmented landscape of almost the same size. The fragments consisted of secondary forest of approximately 80 years of age. Because of the similarities in age and vegetation structure, the area offered optimal opportunities to investigate the effects of fragmentation on the ecology of the focus species. Five fragments of different sizes and the continuous forest area were included in the analyses.

Firstly, I investigated the hypothesis that specialist species in contrast to generalists are possibly restricted in their movement distances in smaller fragments compared to larger fragments. In this investigation two other rodent species, *Oryzomys russatus* (Rice rat) and the endemic *Thaptomys nigrita* (Blackish grass mouse), were included because of the scarce knowledge about these species. Results revealed that the generalist rodent *A. montensis* moved the smallest mean distances compared to other species, while the two marsupials moved much larger distances than all rodents, implying larger home ranges. Mean maximum distances moved were also smallest for *A. montensis* and largest for the two marsupial species. All rodent species moved most frequently short distances, and more than 80% of their movements did not exceed 50 m. With respect to the hypothesis no effects of fragmentation were detected since distances moved did not differ between study sites for any of the investigated species. Obviously, movement distances are not impaired by fragmentation and even the smallest investigated fragments were big enough to meet the spatial demands of the focus species. *O. nigripes*, *O. russatus*, and *G. microtarsus* were the only species showing differences in mean distances moved between sexes with females moving lower distances. Males of *G. microtarsus* showed a seasonal variation in mean distances moved, moving longer distances in the rainy season, probably due to reproductive activity.

In the second chapter I investigated the area-density relationship of the focus species. I hypothesized that specialist species should decrease in population density with decreasing fragment size while the density of common generalist species should be unaffected or even increase in smaller forest remnants. Population densities of the focus species were achieved and correlated to four fragmentation variables (size of the fragment, amount of edge between forest and matrix, connectivity of the fragment, and proportion of anthropogenic altered area around the fragment). In the results distinct differences between species became apparent. Both generalist rodent species *A. montensis* and *O. nigripes* showed no correlation to any of the landscape variables, suggesting that fragmentation effects do not influence densities of these species. However, densities of both species varied between fragments, implying other factors not investigated to be important. On the other hand, density of the specialized rodent *D. sublineatus* was positively correlated to size and negatively to amount of forest edge of the fragments. Possible reasons for this negative relationship between density and fragmentation are decreased immigration into smaller fragments, decreased juvenile recruitment in smaller compared to larger fragments and/or increased mortality due to edge effects in smaller fragments with increased relative amount of habitat edge. Density of the more generalist marsupial *G. microtarsus* did not show any relation to the fragmentation variables. Surprisingly, density of *M. incanus* was negatively correlated to connectivity of the fragment. Reasons for this remain obscure, especially because the result is contradictory to other investigations. Possibly, other effects that influence population density such as for example food availability, put fragmentation effects into perspective in this species.

Fragmentation can influence vegetation structures for example by increased edge effects in smaller fragments. The vegetation structures of a forest can determine the habitat suitability for certain species. Therefore, I examined in the third chapter the response of the focus species to different vegetation structures. The hypothesis was that for specialist species mature forest characteristics are obligatory while generalists are correlated to disturbed forest characteristics. Results proved that

changes in the structure of the forest might influence small mammal species in their spatial distribution. Investigations on the preferred vegetation structures revealed a clear preference for habitat characterized by a low canopy and dense understory vegetation for *A. montensis* and *O. nigripes*. *M. incanus* and *D. sublineatus* were predominantly captured in habitat characterized by a closed canopy and *G. microtarsus* in habitat characterized by open canopy, respectively. These results lead to the conclusion that not only obvious changes of habitat due to fragmentation like fragment size reduction but also collateral structural changes in the forest have consequences for inhabiting species. Further, the strength of preference was negatively correlated with fragment size and positively with the availability of preferred vegetation structure in generalists *A. montensis* and *O. nigripes*. This was not due to simply more preferred vegetation structure in the smaller fragments. That means that some other unknown effect is taking place, probably a competitive release for these species due to fewer species in the smaller fragments. This was however not tested here.

Parasites play an important role in natural communities and various studies indicated that they are able to control host populations in size and demography analogue to the impact of predators or limitation of resources. In this context, I investigated the effect of fragmentation on parasite load and body condition of the focus species. I hypothesized that body condition decreases with increasing parasite load and that parasite burden increases with increasing fragmentation in specialist species but not in generalist species as a consequence of differing responses to fragmentation effects. Results revealed a high prevalence in all species except for the arboreal *G. microtarsus* presumably because of decreased infection probability. No correlation was found between body condition and parasite load in any of the species. With respect to the hypothesis, body condition of specialist species was more influenced by fragmentation compared to that of generalists, but in a way that was unexpected. Body condition turned out to be positively correlated with fragmentation effects in specialists *D. sublineatus* and *M. incanus*. Further, parasite load increased in *D. sublineatus* with population density, probably due to improved contact rates and therefore transmission of parasites between congeners. Fragmentation has obviously an indirect effect on parasite load in this species by affecting population densities. Probably lower parasite burden in smaller fragments augments the effect because individuals with fewer parasites are able to maintain a higher body condition. For generalist species low or no correlation between parasite burden and fragmentation was detected suggesting little effects of fragmentation on these species.

The results of this study revealed different effects of fragmentation on ecological parameters of specialist and generalist small mammal species of the Atlantic forest and contribute to the collection of information on these poorly studied species. Information like this is needed to identify the demands of the species in order to identify the species most vulnerable to habitat fragmentation. The findings underline the importance of large forest fragments for the persistence of endemic specialized small mammal species in the Atlantic forest. Only these large fragments and continuous forest parts can provide habitat and vegetation structures which are able to support persisting populations of endangered species like *D. sublineatus*.

ZUSAMMENFASSUNG

Die Fragmentierung von Lebensraum wird als Hauptbedrohung für die Biodiversität weltweit und insbesondere in tropischen Regenwäldern angesehen. Der Atlantische Küstenregenwald Brasiliens (Mata Atlântica) ist einer der 25 Biodiversitäts-Hotspots der Erde und gleichzeitig einer der weltweit meist bedrohten Lebensräume. Er ist die Heimat von über 20000 Pflanzen und über 1360 Wirbeltierarten. Zusätzlich kennzeichnet sein hoher Grad an endemischen Arten (ca. 40 % der Pflanzen sowie der Vertebraten sind auf diesen Lebensraum begrenzt) den hohen Wert des Atlantischen Küstenregenwaldes für den Naturschutz.

Kleinsäuger (Säugetiere unter einem kg Körpergewicht, ausgenommen Affen und Fledermäuse) reagieren auf verschiedene Weise auf Fragmentierung des Habitats Wald. Habitatgeneralisten werden - im Gegensatz zu den Spezialisten – von der Fragmentierung des Lebensraumes nicht oder kaum eingeschränkt. Die Gründe für diese Unterschiede zwischen Kleinsäugern des Atlantischen Küstenregenwaldes sind kaum bekannt und mit meinen Untersuchungen versuche ich das Wissen über die Ökologie dieser Arten zu erweitern. Ich habe den Effekt von Fragmentierung auf die Bewegungsmuster, die Populationsdichte, die Nutzung von Vegetationsstrukturen, und die Parasitenbelastung von Kleinsäugern des Atlantischen Küstenregenwaldes in der Nähe der Stadt São Paulo, Brasilien, untersucht. Die zugrunde liegende Frage in allen Untersuchungen war: Werden diese Parameter der Kleinsäugerökologie von Habitatfragmentierung beeinflusst?

Aufgrund vorhergehender Untersuchungen unserer brasilianischen Kooperationspartner wurden Fokusarten ausgesucht (Nagetiere: *Akodon montensis* (Grass mouse), *Oligoryzomys nigripes* (Black-footed pygmy rice rat) und die endemische Art *Delomys sublineatus* (Pallid Atlantic forest rat); Beuteltiere: *Marmosops incanus* (Gray slender mouse opossum) und *Gracilinanus microtarsus* (Brazilian gracile mouse opossum). Beide Beuteltierarten sind endemisch im Atlantischen Küstenregenwald. *A. montensis*, *O. nigripes* und *G. microtarsus* reagieren eher als Generalisten in Bezug auf die Fragmentierung von Lebensraum, während *M. incanus* und *D. sublineatus* eher spezialisierte Waldarten sind, die empfindlich auf Waldfragmentierung reagieren.

Das Untersuchungsgebiet bestand aus einem zusammenhängenden Waldgebiet von ca. 10 000 ha Größe und einer ungefähr gleich großen benachbarten fragmentierten Landschaft. Alle Waldfragmente bestanden aus Sekundärwald mit einem Alter von etwa 80 Jahren. Durch die Ähnlichkeit der Waldfragmente in Alter und Vegetationsstruktur bot das Untersuchungsgebiet optimale Voraussetzungen für die Untersuchung von Fragmentierungseffekten auf die Ökologie der Fokusarten. Fünf Waldfragmente unterschiedlicher Größe wurden in die Untersuchung einbezogen.

Zuerst untersuchte ich die Hypothese, dass Spezialisten im Gegensatz zu Generalisten möglicherweise in kleinen Fragmenten stärker in ihren Bewegungsmustern eingeschränkt werden als in großen Fragmenten. In diese Untersuchung wurden zwei weitere Nagerarten, *Oryzomys russatus* (Rice rat) und die endemische Art *Thaptomys nigrita* (Blackish grass mouse), einbezogen. Die Ergebnisse zeigten, dass der generalistische Nager *A. montensis* im Vergleich mit den anderen untersuchten Arten die geringsten mittleren Entfernungen zurücklegte, während die beiden Mausopossums größere Entfernungen als alle Nager zurücklegten, was auf einen größeren Aktionsradius hindeutet. Alle Nager bewegten sich am häufigsten über kurze Distanzen und mehr als 80 % ihrer Bewegungen waren nicht länger als 50 m. In Bezug auf die Hypothese wurden keine Effekte von Fragmentierung gefunden, da sich die mittlere zurückgelegte Entfernung für keine der untersuchten Arten zwischen den Fragmenten unterschied. Offenbar werden die zurückgelegten Entfernungen nicht von Fragmentierungseffekten beeinflusst und auch die kleinen untersuchten Fragmente waren groß genug, um den räumlichen Ansprüchen dieser Kleinsäu

ger zu entsprechen. Bei *O. nigripes*, *O. russatus* und *G. microtarsus* unterschieden sich die mittleren zurückgelegten Entfernungen zwischen den Geschlechtern, wobei jeweils die Weibchen die kürzeren Distanzen zurücklegten. Weiterhin zeigte sich bei den Männchen von *G. microtarsus* saisonale Variation in der mittleren zurückgelegten Entfernung mit einer Steigerung am Ende der Regenzeit, was eventuell auf Paarungsaktivität zurückzuführen ist.

Im zweiten Kapitel untersuchte ich das Flächen/Dichte-Verhältnis der Fokusarten. Ich hypothetisierte, dass die Populationsdichte der Spezialisten mit abnehmender Fragmentgröße abnimmt, während die Populationsdichte der Generalisten von der Fragmentgröße unbeeinflusst bleibt oder sogar in kleinen Fragmenten zunimmt. Populationsdichten der Fokusarten wurden ermittelt und mit den vier Fragmentierungsparametern Fragmentgröße, Länge des Waldrandes zwischen Wald und Matrix, Verbindung zu anderen Fragmenten und Anteil an anthropogen veränderter Fläche um das Fragment korreliert. Die Ergebnisse zeigten klare Unterschiede zwischen den Arten. Beide generalistischen Nagerarten *A. montensis* und *O. nigripes* zeigten keine Korrelation zu einer der Fragmentierungsparametern, was darauf hindeutet, dass Fragmentierungseffekte keinen Einfluss auf die Populationsdichten dieser Arten hat. Allerdings schei nen andere Faktoren, die hier nicht untersucht wurden, einen wichtigen Einfluss auf die Populationsdichten zu haben, da die Dichte beider Arten zum Teil stark zwischen den Fragmenten variierte. Die Populationsdichte des Spezialisten *D. sublineatus* war sowohl positiv mit der Größe des Fragments als auch negativ mit der Länge des Waldrandes korreliert. Mögliche Ursachen für diesen negativen Zusammenhang zwischen Populationsdichte und Fragmentierung sind verringerte Immigration in kleineren Fragmenten, verringerter Populationszuwachs durch Geburten in kleinen gegenüber großen Fragmenten und/oder erhöhte Mortalität durch Effekte, die durch das vergrößerte Verhältnis von Fragmentgröße zu Waldrand vermehrt in kleineren Fragmenten auftreten. Die Populationsdichte der eher generalistischen Beuteltierart *G. microtarsus* war mit keinem der Fragmentierungsparametern korreliert. Überraschenderweise zeigte die Dichte von *M. incanus* eine negative Relation mit der Verbundenheit zu anderen Fragmenten. Die Gründe für dieses Ergebnis bleiben unklar, insbesondere da sie vorherigen Untersuchungen widersprechen. Möglicherweise werden in diesem Fall die Effekte von Fragmentierung von anderen Einflüssen auf die Populationsdichte, wie zum Beispiel die Verfügbarkeit von Nahrung, überdeckt.

Fragmentierung kann die Vegetationsstrukturen, z. B. durch verstärkte Randeffekte in kleineren Fragmenten, beeinflussen. Die Vegetationsstruktur eines Waldes kann wiederum die Eignung des Habitats für bestimmte Arten bestimmen. Im dritten Kapitel habe ich daher die Reaktion der Fokusarten auf unterschiedliche Vegetationsstrukturen untersucht. Die Hypothese war, dass für Spezialisten Strukturen eines älteren Waldes obligat sind, während Generalisten eher mit Merkmalen eines gestörten Waldes korreliert sind. Die Ergebnisse zeigten, das Veränderungen der Vegetationsstruktur des Waldes, die durch Fragmentierung hervorgerufen werden können, einen Einfluss auf die räumliche Verteilung von Kleinsäugern haben können. Untersuchungen bezüglich der präferierten Vegetationsstruktur ergaben bei den Generalisten *A. montensis* und *O. nigripes* eine klare Vorliebe für Habitat, das durch niedriges Kronendach und dichte Unterholzvegetation charakterisiert ist. *M. incanus* und *D. sublineatus* hingegen wurden bevorzugt in Habitat gefangen, das sich durch dichtes Kronendach auszeichnet, während *G. microtarsus* Gebiete mit offenem Kronendach bevorzugte. Diese Ergebnisse lassen den Schluss zu, dass nicht nur offensichtlich Effekte von Fragmentierung wie z. B. die Verkleinerung von Fragmenten, sondern auch begleitende Waldstrukturänderungen Konsequenzen für die dort lebenden Arten haben. Zusätzlich war bei *A. montensis* und *O. nigripes* die Ausprägung der Präferenz für niedriges Kronendach und dichtes Unterholz negativ mit der Fragmentgröße korreliert, was nicht dadurch erklärt werden konnte, dass in den kleinen Fragmenten mehr präferiertes Habitat zu finden war. Das bedeutet, dass ein anderer Einfluss besteht, eventuell eine Konkurrenzvermeidung, da in kleineren Fragmenten weniger Arten vorhanden sind. Dies bedarf jedoch weitergehender Untersuchungen.

Parasiten spielen in natürlichen Gemeinschaften eine wichtige Rolle und verschiedene Studien zeigen, dass sie in der Lage sind, Wirtspopulationen analog zu Prädatoren oder limitierten Ressourcen in Größe und Demografie zu kontrollieren. In diesem Knontext habe ich die Kondition und die Parasitenbelastung der Fokusarten in Korrelation mit Fragmentierungseffekten untersucht. Ich untersuchte die Hypothese, dass die Kondition mit zunehmender Parasitenbelastung abnimmt und die Parasitenbelastung der Spezialisten - im Gegensatz zu der Belastung der Generalisten - mit zunehmender Fragmentierung zunimmt. Die Prävalenz war bei allen Arten bis auf *G. microtarsus* hoch, was wahrscheinlich auf die arboreale Lebensweise dieser Art und die damit verbundene Verminderung des Infektionsrisikos zurückzuführen ist. Bei keiner der untersuchten Arten wurde ein Zusammenhang zwischen Kondition und Parasitenbelastung festgestellt. Die Hypothese, dass Kondition von Spezialistenarten verglichen zu

Generalistenarten stärker beeinflusst wird, konnte bestätigt werden, allerdings in unerwarteter Weise. Ich erwartete, dass Spezialisten in größeren Fragmenten höhere Kondition aufweisen, doch es zeigte sich, dass bei *D. sublineatus* und *M. incanus* die Kondition positiv mit Fragmentierungseffekten korreliert war. Weiterhin stieg die Parasitenbelastung bei *D. sublineatus* mit der Populationsdichte, wahrscheinlich verursacht durch erhöhte innerartliche Kontaktraten und dadurch verbesserte Übertragungsbedingungen für Parasiten. Habitatfragmentierung hat bei dieser Art offensichtlich einen indirekten Effekt auf die Parasitenbelastung durch den Einfluss auf die Populationsdichte. Eventuell wird der Effekt dadurch verstärkt, dass Individuen mit geringerer Parasitenbelastung in kleineren Fragmenten besser in der Lage sind, eine gute Kondition aufrecht zu erhalten. Für Generalistenarten wurde geringe oder gar keine Korrelation zwischen Parasitenbelastung und Fragmentierungsparametern gefunden, was darauf hinweist, dass Habitatfragmentierung einen geringen Einfluss bei diesen Arten hat.

Die Ergebnisse dieser Studie zeigen verschiedene Effekte von Fragmentierung auf spezialisierte und generalistische Kleinsäugerarten des Atlantischen Küstenregenwaldes und leisten einen Beitrag zu der Sammlung von Informationen über diese wenig untersuchten Kleinsäugerarten. Informationen dieser Art sind notwendig, um die am meisten durch Habitatfragmentierung bedrohten Arten und deren Ansprüche an das Habitat zu identifizieren. Die Ergebnisse unterstreichen die wichtige Rolle von großen Waldfragmenten und zusammenhängendem Restwald für das Bestehen von endemischen, spe

zialisierten Kleinsäugerarten im Atlantischen Küstenregenwald. Nur diese großen Fragmente sind in der Lage, den Lebensraum und die Vegetationsstruktur zu bieten, die ein langfristiges Bestehen von Populationen gefährdeter Arten wie *D. sublineatus* ermöglichen.

GENERAL INTRODUCTION

A GLOBAL PROBLEM FOR BIODIVERSITY – HABITAT FRAGMENTATION

Habitat fragmentation is considered as a serious threat to global biodiversity (Saunders *et al.* 1991, Terborgh 1992, Andrén 1994, Laurance 1994, Laurance and Cochrane 2001, Tabarelli *et al.* 2004). Fragmentation of a landscape is characterized as a process during which a large continuous habitat is converted into a series of smaller remnants separated by a matrix of different vegetation. It is for the most part due to the anthropogenic removal of natural vegetation which is replaced by agricultural or urban areas. The implications of landscape fragmentation are complex. The vegetation remnants may differ in size, microclimate (within the remnant as well as of the surrounding matrix), isolation, shape, neighbouring matrix or biogeographic aspects. Additionally, the time since isolation may be different between fragments of an originally connected habitat (Saunders et al. 1991). Habitat fragmentation involves both habitat loss and fragmentation per se, the breaking apart of habitat (Laurance and Cochrane 2001). Compared to fragmentation, habitat loss is considered to have greater, consistently negative

effects on biodiversity (Fahrig 2002, Fahrig 2003).

Several studies showed that habitat loss may lead to reduced trophic chain length and reduced number of specialist species, effects breeding and dispersal success as well as predation rate and further influences animal behaviour and species interaction (reviewed in Fahrig 2003). Although fragmentation of habitat is in most cases inextricably linked to habitat loss, fragmentation per se may have varying effects on species diversity or ecologic processes. Different ecosystems and organisms react in different or even contradictory ways on fragmentation of the habitat (Robinson *et al.* 1992, Haila 2002). A severe effect of fragmentation is the dramatic increase of habitat edge in relation to habitat area (Laurance 2000). This results in various changes and alterations of abiotic factors in the involved part of the habitat (edge effects). In forest ecosystems, edge effects include increased solar radiation, increased desiccation and temperature variability near forest edges (Saunders *et al.* 1991, Stevens and Husband 1998). Increased wind effects along forest edges may cause structural damage in the forest, and there is evidence that seed dispersal and nutrient cycling is altered by edge effects (Bierregaard *et al.* 1992). Further, an increased amount of forest edges can make it easier for exotic species to invade habitat patches (Debinski and Holt 2000, Ganzhorn 2003). Several investigations documented the negative effect of fragmentation on a variety of different plant and animal taxa. For example, Burkey (1995) showed that extinction probabilities of lizards, birds and small mammals were higher for a set of small fragments compared to a single large fragment of the same area given the fragments are isolated from each other. Forest fragments in Amazonia harboured only a subset of ant species found in continuous forest, and the number of rare ant species also declined in forest fragments (Vasconcelos *et al.* 2006). Diffendorfer *et al.* (1995) found reduced movement rates and altered spatial patterning of distances moved in small mammal species in a fragmented habitat and many studies demonstrated increased nest predation and therefore reduced nesting success of songbirds near forest edges (review Hartley and Hunter 1998).

On the other hand, fragmentation might have positive effects on species or biodiversity. For example, Laurance *et al.* (1998) found increased recruitment rates in trees along forest edges compared to interior forest in Amazonia. Adult female voles had larger home ranges, body sizes, residence times and reproductive rates along habitat edges compared to individuals in the interior of a habitat patch (Bowers and Matter 1997). Tscharntke *et*

al. (2002) showed that many small fragments of calcareous grassland supported a higher percentage of overall butterfly species richness than one or two big fragments of the same area even when only endangered butterfly species were considered.

These studies showed that different organisms react in different ways to the fragmentation of habitat. The effects of fragmentation differ not only between species but also between habitat types and geographic regions (Haila 2002). However, often accompanied with fragmentation is the anthropogenic land use of the matrix which may affect remaining habitat patches. Forest remnants often are selectively logged, degraded by groundfires or overhunted (Laurance 2001, Laurance and Cochrane 2001, Tabarelli *et al.* 2004). Cattle grazing, hydrological changes and the use of pesticides and herbicides in agricultural and urban areas can also have dramatic consequences for the ecology of fragment inhabiting species (Myers 1987, 1989, Hobbs and Huenneke 1992). Considering these effects fragmentation and habitat loss may lead to other probable harmful changes which are induced by human activity inside the remaining fragments or in the matrix between them (Haila 2002).

The vulnerability of a species to fragmentation depends strongly on its ability to deal with such alterations and its capability to use the surrounding matrix (Gascon *et al.* 1999). Moreover, fragmentation effects depend on home range boundaries and movement patterns of the individual species (Haila 1990, Haila *et al.* 1993). The assumption that fragmentation divides a continuous habitat in habitat patches separated by hostile vegetation is not true for generalist species, for which a fragmented landscape rather represents a continuous habitat with different suitability (Andrén 1994). Habitat generalists may be able to survive even in very small remnants because of their ability to benefit from the resources in the surrounding matrix. Further, species diversity may increase after fragmentation of a continuous habitat due to generalist species which can be found in the newly created anthropogenic habitat (Andrén 1994). For example, Tocher *et al.* (1997) found an increase in frog species richness after fragmentation in a forest habitat caused by immigration of generalist species from the surrounding matrix.

On the other hand, specialists which depend on a certain habitat and which are influenced by edge effects are most vulnerable to fragmentation. Specialist species are less able to use resources different to the original habitat.

Another possible effect of fragmentation is its probable impact on the distribution of parasites and diseases. The dynamic of parasites strongly depends on the density of the host species because of induced stress (May and Anderson 1979) and improved transmission circumstances at high host densities. Furthermore, the dispersal ability of the host is an important parameter in parasite distribution. Generalist species, capable of occupying a range of habitats, may act as "bridge" species, facilitating the transmission of parasites among fragments. In specialist species, fragmentation may lead to increased environmental stress and the individual may be faced with nutritional problems. In the case where overall body condition of an individual is diminished due to fragmentation effects (Díaz *et al.* 1999), the individual will be more vulnerable to diseases or parasites. This can further reduce body condition and additionally impair the individual. If host populations are isolated, pathogens or parasites can have a severe impact on these populations (e. g. Macdonald 1996, Allan *et al.* 2003).

For these reasons, successful conservation requires the identification of the specialist and most vulnerable species in the given region and an estimate of the minimum habitat required by that species (Debinski and Holt 2000, Fahrig 2003). Knowledge about range of habitat and factors limiting populations is indispensable and allows predictions about fragmentation effects on the species to be made. This specific information about the response of certain species is essential to concentrating conservation efforts on the requirements needed by especially endangered species.

In this thesis, I examined small mammals in a neotropical rainforest habitat. Tropical rainforests are one of the most diverse ecosystems in the world and are extremely threatened by habitat fragmentation and human disturbance (Turner 1996, Tabarelli *et al.* 2004). Although tropical rainforests cover only about 7% of the land mass they are inhabited by half to two-thirds of the species of plants and animals on earth (Bierregaard *et al.* 1992). This underlines the importance of research on fragmentation and resulting conservation efforts.

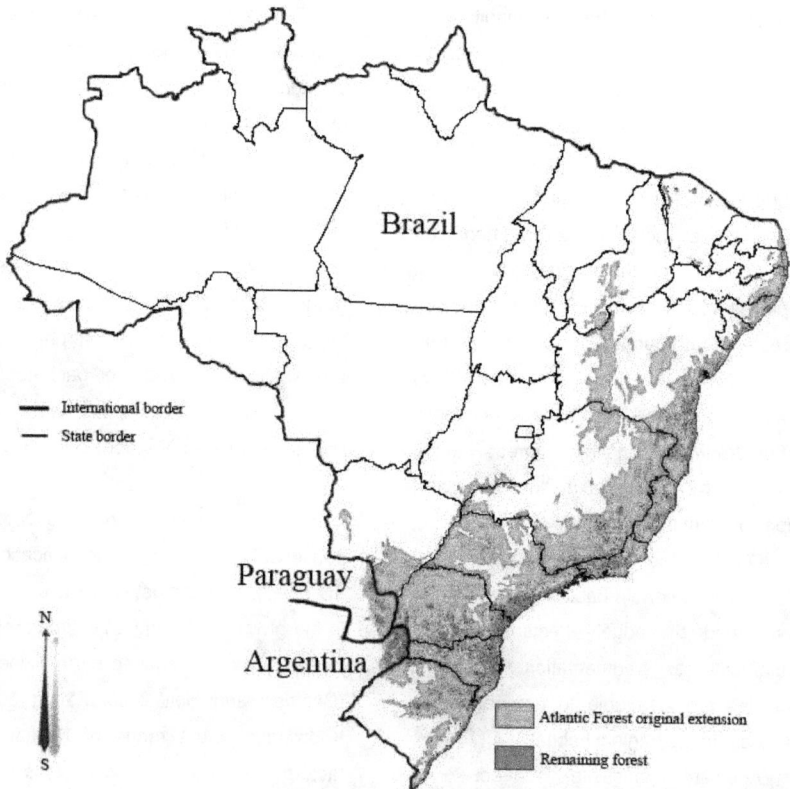

Fig. 1 Original and actual extension of the Atlantic Forest (modified from Galindo-Leal and de Gusmão Câmara 2003)

BIODIVERSITY HOTSPOT –
MATA ATLÂNTICA, BRAZIL

Among the earth´s rainforest habitats, the coastal Atlantic forest (Mata Atlântica) in Brazil is one of the 25 biodiversity hotspots in the world (Myers *et al.* 2000) and was assigned as a Biosphere Reserve by the UNESCO in 1992. Despite its high degree of destruction the Atlantic forest is home to over 20,000 plant species and more than 1,360 vertebrate species. Additionally, its high degree of endemism justifies its value for biodiversity conservation. 181 birds, 73 mammal species, 60 reptiles and 253 amphibian species are restricted to this biome, which equals about 41 % of the vertebrate species in the Atlantic forest and 2.1 % of global vertebrates. Furthermore, about 8000 plant species (40 % of inhabiting species, 2.7 % of global plants) are endemic to the Atlantic forest (Myers *et al.* 2000). However, currently at least 510 plant and vertebrate species of the Atlantic forest are officially threatened (Tabarelli *et al.* 2003). The Atlantic forest originally covered an area of approximately 1 to 1.5 million square kilometers from the state of Rio Grande do Norte in the north to Rio Grande do Sul in the south of Brazil as well as parts of northeastern Argentina and eastern Paraguay (Galindo-Leal and de Gusmão Câmara 2003; Fig. 1). Due to intensive deforestation and land use (wood exploitation, sugarcane plantation, coffee, cacao, cattle grazing, eucalyptus and pine plantation; Fig. 2) in the last few

Fig. 2 Anthropogenic pressure on the Atlantic forest. Upper row from left to right: agricultural used matrix between forest fragments near São Paulo; harvested sugarcane plantation adjacent to remaining forest, Recife (with courtesy of D. Piechowski); stubbed and burned *Eucalyptus* plantation close to forest fragments near São Paulo; below: skyline of São Paulo

centuries only about 7 % of the original forest cover remains, which does not only include primary forest, but also secondary forest remnants in different stages of regeneration (Galindo-Leal & Câmara 2003, Dunn 2004). The stage of regeneration of a forest and the degree of fragmentation influence the forest structure, which can determine habitat suitability for certain species and affect their occurrence as well as the composition of animal communities (Tews et al. 2004). Thus, secondary forests may play an important role in species conservation if primary forest habitats are limited (Dunn 2004). The remaining forest is highly fragmented (Ranta et al. 1998). Today the coastal Atlantic forest extends from 4° to 32° S and still covers a wide range of climatic belts and vegetation formations. In the northeastern part of Brazil it occupies a thin coastal strip not exceeding 40 miles in width, while in the south it extends from the coast to as far as 200 miles inland. Elevation ranges from sea level to 2900 m.

Deforestation still goes on and has been especially severe in the last decades. 11,650 km^2 have been lost during the last 15 years (Tabarelli et al. 2005). More than 100 million people live in the area of the Atlantic forest in Brazil (Galindo-Leal and de Gusmão Câmara 2003), especially concentrated in the four urban centers of São Paulo, Rio de Janeiro, Curitiba and Belo Horizonte, where the population has increased substantially in the last 50 years. This has led to further anthropogenic pressure through uncontrolled urban expansion and industrialization.

Besides its importance for biodiversity, the forest also plays a fundamental role for the 70 % of the Brazilian people inhabiting the coastal region of Brazil. It acts as a water reservoir and maintains the soil fertility which is important for food production. Furthermore, increased deforestation has been accompanied by reduced job opportunities in the Atlantic forest areas converted to agricultural use in the southern states of Brazil (Frickmann Young 2003). Therefore, management plans for the remaining Atlantic forest are needed which not not only focuse on nature conservation but also on the requirements of local human populations. This is necessary to ensure successful conservation of the ecosystem as a whole.

ECOLOGICALLY IMPORTANT FOREST INHABITANTS – SMALL MAMMALS

Non-volant small mammals are defined as all mammals not exceeding 1 kg in weight excluding primates and bats. Small mammals are the most diverse ecological group within the neotropical mammals and more than 200 species are recorded in

Brazil (Fonseca *et al.* 1996). In the Atlantic forest 92 species of small mammals are known of which 43 are endemic (Fonseca *et al.* 1996). Small mammals play a key role in forest ecosystems in terms of seed dispersal (Forget 1991, Sánchez-Cordero and Martínez-Gallardo 1998, Brewer and Reimánek 1999, Vieira and Izar 1999, Vieira *et al.* 2003, Pimentel and Tabarelli 2004) and the dispersal of mycorrhizal fungi (Janos and Sahley 1995). Furthermore, forest regeneration is strongly effected by small mammals through predation on seedlings (Forget 1993, Pizo 1997, Forget *et al.* 2000, Vieira *et al.* 2003) and changes in their abundance have been shown to effect forest regeneration and succession (Terborgh *et al.* 2001). The important ecological function of small mammals is disproportionate to the knowledge about the ecology of small mammal species in the Atlantic forest (Fonseca and Kierulff 1989, Barros-Battesti *et al.* 2000). Despite the high species number, basic scientific knowledge on taxonomy, systematics, distribution and natural history of small mammals in Brazil is limited which in turn is one of the major threats to this group (Costa *et al.* 2005). However, several investigations in the Atlantic forest indicated that small mammals respond to habitat fragmentation and the accompanied alteration in forest structure (Stallings 1989, Fonseca and Robinson

1990, Pires *et al.* 2002, Pardini 2004, Pardini *et al.* 2005).

BMBF PROGRAM – SCIENCE AND TECHNOLOGY FOR THE MATA ATLÂNTICA

This study is part of the cooperation project MATA ATLÂNTICA - Science and Technology for the Mata Atlântica - of the German Ministry of Science and Education (BMBF, Bundesministerium für Forschung und Bildung) and the Brazilian National Council for Scientific and Technological Development (CNPq, Conselho Nacional de Desenvolvimento Científico e Tecnológico). The project was launched in 2002 and aims to develop strategies and action plans for the conservation, sustainable management and use of the Brazilian Atlantic forest. It comprises a total of five collaborating sub-projects with Brazilian counterparts and study areas in Recife (Pernambuco, PE), Rio de Janeiro (Rio de Janeiro, RJ), São Paulo (São Paulo, SP), Curitiba (Parana, PR) and Florianópolis (Santa Catarina, SC).

The work is part of the São Paulo sub-project BioCAPSP (Biodiversity Conservation in fragmented landscapes at the Atlantic Plateau of São Paulo) in co-operation with the University of São Paulo. The interdisciplinary project aims to investigate small mammals of the Atlantic forest on population, community and individual level as well as influences of fragmentation on immunocompetence and parasite load of the species.

GENERALIST VS. SPECIALIST SPECIES

In the Atlantic forest the group of small mammals comprises rodent and marsupial species. Our main focus was to investigate the influence of fragmentation on generalist as well as specialist small mammal species original inhabiting the Atlantic forest and to collect a reasonable amount of data in order to be able to generalize findings about these species. Based on previous studies by our Brazilian collaborator Dr. Renata Pardini and her students (Pardini et al. 2005, Umetsu 2005, Umetsu and Pardini in press), I focused mainly on three rodent and two marsupial species. One of the rodent species and both marsupials are endemic to the Atlantic forest while the other species occur in other biomes as well. The focus species can be separated in habitat generalists who are - in contrast to specialist species - not impaired by fragmentation of the forest habitat. These differences became apparent for example by different capture rates of species in forest fragments of different sizes. The reasons for these differences were unknown and my investigations aimed to emphasize specific knowledge about the ecology of these species in order to provide detailed data for effective conservation planning. I concentrated on four ecological aspects. Firstly, I investigated the movement patterns as well as maximum movement distances of the species in order to identify possible restrictions of spatial requirements of the focus species caused by fragmentation (Chapter 1). Secondly, I examined the area-density relationship of the species by investigating the abundances of the focus species in forest fragments of different sizes (Chapter 2). Further, I analyzed specific habitat preferences of the focus species. Thereby, I intended to identify major traits of vegetation structures or forest features that are mandatory for the species (Chapter 3). Finally, I examined the body condition and the parasite burden of the focus species in order to figure out possible effects of habitat fragmentation on the conditional status of the small mammals (Chapter 4). With respect to these aspects, I concentrated on the four main questions:

1. Does fragmentation influence the movement distances of the species?
2. What are the area-density-relationships of the species?
3. Do the species prefer distinct forest vegetation structures?

and, finally

4. Is parasite burden influenced by habitat fragmentation?

With these questions I tried to figure out the difference in response to habitat fragmentation between generalist and specialist rodent and marsupial species and to identify the minimum requirements of the specialist species in a tropical rainforest.

REFERENCES

Allan, B. F., F. Keesing, and R. S. Ostfeld. 2003. Effect of forest fragmentation on Lyme disease risk. Conservation Biology 17:267-272.

Andrén, H. 1994. Effects of habitat fragmentation on birds and mammals in landscapes with different proportions of suitable habitat: a review. Oikos 71:355-366.

Barros-Battesti, D. M., R. Martins, C. R. Bertim, N. H. Yoshinari, V. L. N. Bonoldi, E. P. Leon, M. Miretzki, and T. T. S. Schumaker. 2000. Land fauna composition of small mammals of a fragment of Atlantic forest in the state of Sao Paulo, Brazil. Revista Brasileira de Zoociencias 17:241-249.

Bierregaard, R. O., T. E. Lovejoy, K. Valerie, A. A. dos Santos, and R. W. Hutchings. 1992. The biological dynamics of tropical Rainforest fragments: a prospective comparison of fragments and continuous forest. BioScience 42:859-866.

Bowers, M. A., and S. F. Matter. 1997. Landscape Ecology of mammals: relationships between density and patch size. Journal of Mammalogy 78:999-1013.

Brewer, S., and M. Reimánek. 1999. Small rodents as significant dispersers of tree seeds in a Neotropical forest. Journal of Vegetation Science 10:165-174.

Burkey, T. V. 1995. Extinction rates in archipelagoes: Implications for populations in fragmented habitats. Conservation Biology 9:527-541.

Costa, L. P., Y. L. R. Leite, S. L. Mendes, and A. D. Ditchfield. 2005. Mammal conservation in Brazil. Conservation Biology 19:672-679.

Debinski, D. M., and R. D. Holt. 2000. A survey and overview of habitat fragmentation experiments. Conservation Biology 14:342-355.

Díaz, M., T. Santos, and J. L. Tellería. 1999. Effects of forest fragmentation on the winter body condition and population parameters of an habitat generalist, the wood mouse *Apodemus sylvaticus*: a test of hypotheses. Acta Oecologica 20:39-49.

Diffendorfer, J. E., M. S. Gaines, and R. D. Holt. 1995. Habitat fragmentation and movements of three small mammals (*Sigmodon, Microtus,* and *Peromyscus*). Ecology 76:827-839.

Dunn, R. R. 2004. Recovery of faunal communities during tropical forest regeneration. Conservation Biology 18:302-309.

Fahrig, L. 2002. Effect of habitat fragmentation on the extinction threshold: a synthesis. Ecological Applications 12:346-353.

Fahrig, L. 2003. Effects of habitat fragmentation on biodiversity. Annual Review of Ecology, Evolution and Systematic 34:487-515.

Fonseca, G. A. B., G. Herrmann, Y. L. R. Leite, R. A. Mittermeier, A. B. Rylands, and J. L. Patton. 1996. Lista anotada dos mamíferos do

Brasil. Occasional Papers in Conservation Biology 4:1-38.

Fonseca, G. A. B., and M. C. M. Kierulff. 1989. Biology and natural history of brazilian Atlantic forest small mammals. Bulletin of the Florida State Museum, Biological Sciences 34:99-152.

Fonseca, G. A. B., and J. G. Robinson. 1990. Forest size and stucture: competitive and predator effects on small mammal communities. Biological Conservation 53:265-294.

Forget, P.-M. 1991. Evidence for secondary seed dispersal by rodents in Panama. Oecologia 87:596-599.

Forget, P.-M. 1993. Post-dispersal predation and scatterhoarding of *Dipteryx panamensis* (Papilionaceae) seeds by rodents in Panama. Oecologia 94:255-261.

Forget, P.-M., T. Milleron, F. Feer, O. Henry, and G. Dubost. 2000. Effects of dispersal pattern and mammalian herbivores on seedling recruitment for *Virola mechelii* (Myristicaceae) in French Guiana. Biotropica 32:452-462.

Frickmann Young, C. E. 2003. Socioeconomic causes of deforestation in the Atlantic Forest of Brazil. Pages 103-117 *in* C. Galindo-Leal and I. de Gusmão Câmara, editors. The Atlantic Forest of South America: Biodiversity status, threats, and outlook. Island Press, Washington, Covelo, London.

Galindo-Leal, C., and I. de Gusmão Câmara. 2003. The Atlantic forest of South America: Biodiversity status, threats, and outlook. Island Press, Washington, Covelo, London.

Ganzhorn, J. U. 2003. Effects of introduced *Rattus rattus* on endemic small mammals in dry deciduous forest fragments of western Madagascar. Animal Conservation 6:147-157.

Gascon, C., T. E. Lovejoy, R. O. Bierregaard, J. R. Malcolm, P. C. Stouffer, H. L. Vasconcelos, W. F. Laurance, B. L. Zimmerman, M. Tocher, and S. Borges. 1999. Matrix habitat and species richness in tropical forest remnants. Biological Conservation 91:223-229.

Haila, Y. 1990. Toward an ecological definition of an island: a northwest european perspective. Journal of Biogeography 17:561-568.

Haila, Y. 2002. A conceptual genealogy of fragmentation research: from island biogeography to landscape ecology. Ecological Applications 12:321-334.

Haila, Y., I. K. Hanski, and S. Raivio. 1993. Turnover of breeding birds in small forest fragments: the "sampling" colonization hypothesis corroborated. Ecology 74:714-725.

Hartley, M. J., and M. L. Hunter. 1998. A meta-analysis of forest cover, edge effects, and artificial nest predation rates. Conservation Biology 12:465-469.

Hobbs, R. J., and L. F. Huenneke. 1992. Disturbance, diversity, and invasion: implications for conservation. Conservation Biology 6:324-337.

Janos, D. P., and C. T. Sahley. 1995. Rodent dispersal of vesicular-arbuscular mycorrhizal fungi in Amazonian Peru. Ecology 76:1852-1858.

Laurance, W. F. 1994. Rainforest fragmentation and the structure of small mammal communities in tropical Queensland. Biological Conservation 69:23-32.

Laurance, W. F. 2000. Do edge effects occur over large spatial scales? Trends in Ecology and Evolution 15:134-135.

Laurance, W. F. 2001. Tropical logging and human invasions. Conservation Biology 15:4-5.

Laurance, W. F., and M. A. Cochrane. 2001. Synergistic effects in fragmented landscapes. Conservation Biology 15:1488-1489.

Laurance, W. F., L. V. Ferreira, J. M. Rankin de Merona, S. G. Laurance, R. W. Hutchings, and T. E. Lovejoy. 1998. Effects of forest fragmentation on recruitment patterns in Amazonian tree communities. Conservation Biology 12:460-464.

Macdonald, D. W. 1996. Dangerous liaisons and disease. Nature 379:400-401.

May, R. M., and D. R. Anderson. 1979. Population biology of infectious diseases: Part II. Nature 280:455-461.

Myers, N. 1987. The extinction spasm impending: synergisms at work. Conservation Biology 1:14-21.

Myers, N. 1989. Synergistic interactions and environment. BioScience 39:506.

Myers, N., R. A. Mittermeier, C. G. Mittermeier, G. A. B. Fonseca, and J. Kent. 2000. Biodiversity hotspots for conservation priorities. Nature 403:853-858.

Pardini, R. 2004. Effects of forest fragmentation on small mammals in an Atlantic Forest landscape. Biodiversity and Conservation 13:2567-2586.

Pardini, R., S. Marques de Souza, R. Braga-Neto, and J. P. Metzger. 2005. The role of forest structure, fragment size and corridors in maintaining small mammal abundance and diversity in an Atlantic forest landscape. Biological Conservation 124:253-266.

Pimentel, D. S., and M. Tabarelli. 2004. Seed dispersal of the palm Attalea oleifera in a remnant of the brazilian Atlantic forest. Biotropica 36:74-84.

Pires, A. S., P. Koeler Lira, F. A. S. Fernandez, G. M. Schittini, and L. C. Oliveira. 2002. Frequency of movements of small mammals among Atlantic coastal forest fragments in Brazil. Biological Conservation 108:229-237.

Pizo, M. A. 1997. Seed dispersal and predation in two populations of Cabralea canjerana (Meliaceae) in the Atlantic forest of southeastern Brazil. Journal of Tropical Ecology 13:559-578.

Ranta, P., T. Blom, J. Niemelä, E. Joensuu, and M. Siitonen. 1998. The fragmented Atlantic rain forest of Brazil: size, shape and distribution of forest fragments. Biodiversity and Conservation 7:385-403.

Robinson, G. R., R. D. Holt, M. S. Gaines, S. P. Hamburg, M. L. Johnson, H. S. Fitch, and E. A. Martinko. 1992. Diverse and contrasting effects of habitat fragmentation. SCIENCE 257:524-526.

Sánchez-Cordero, V., and R. Martínez-Gallardo. 1998. Postdispersal fruit and seed removal by forest-dwelling rodents in a lowland rainforest in Mexico. Journal of Tropical Ecology 14:139-151.

Saunders, D. A., R. J. Hobbs, and C. R. Margules. 1991. Biological consequences of ecosystem fragmentation: a review. Conservation Biology 5:18-32.

Stallings, J. R. 1989. Small mammal inventories in an eastern brazilian park. Bulletin of the Florida State Museum, Biological Sciences 34:159-200.

Stevens, S. M., and T. P. Husband. 1998. The influence of edge on small mammals: evidence from Brazilian Atlantic forest fragments. Biological Conservation 85:1-8.

Tabarelli, M., J. M. Cardoso da Silva, and C. Gascon. 2004. Forest fragmentation, synergisms and the impoverishment of neotropical forests. Biodiversity and Conservation 13:1419-1425.

Tabarelli, M., J. M. C. da Silva, and C. M. R. Costa. 2003. Endangered species and conservation planning.

Pages 86-94 *in* C. Galindo-Leal and I. de Gusmão Câmara, editors. The Atlantic Forest of South America: Biodiversity status, threats, and outlook. Island Press, Washington, Covelo, London.

Tabarelli, M., L. P. Pinto, J. M. C. Silva, M. Hirota, and L. Bedê. 2005. Challenges and opportunities for biodiversity conservation in the Brazilian Atlantic forest. Conservation Biology 19:695-700.

Terborgh, J. 1992. Maintenance of diversity in tropical forests. Biotropica 24:283-292.

Terborgh, J., L. Lopez, P. Nuñez, M. Rao, G. Shahabuddin, G. Orihuela, M. Riveros, R. Ascanio, G. H. Adler, T. D. Lambert, and L. Balbas. 2001. Ecological meltdown in predator-free forest fragments. SCIENCE 294:1923-1926.

Tocher, M., C. Gascon, and B. L. Zimmerman. 1997. Fragmentation effects on a central Amazonian frog community: a ten-year study. Pages 124-137 *in* W. F. Laurance and R. O. Bierregaard, editors. Tropical forest remnants: ecology, management, and conservation of fragmented communities. University of Chicago Press, Chicago.

Tscharntke, T., I. Steffan-Dewenter, A. Kruess, and C. Thies. 2002. Contribution of small habitat fragments to conservation of insect communities of grassland-cropland landscapes. Ecological Applications 12:354-363.

Turner, I. M. 1996. Species loss in fragments of tropical rain forest: a review of the evidence. Journal of Applied Ecology 33:200-209.

Umetsu, F. 2005. Pequenos mamíferos em um mosaico de habitats remanescentes e antropogênicos: qualidade da matriz e conectividade em uma paisagem fragmentada de Mata Atlântica. Universidade de São Paulo, São Paulo.

Umetsu, F., and R. Pardini. *in press*. Small mammals in a mosaic of forest remnants and anthropogenic habitats - evaluating matrix quality in an Atlantic forest landscape. Landscape Ecology.

Vasconcelos, H. L., J. M. S. Vilhena, W. E. Magnusson, and A. L. K. M. Albernaz. 2006. Long-term effects of forest fragmentation on Amazonian ant communities. Journal of Biogeography 33:1348-1356.

Vieira, E. M., and P. Izar. 1999. Interactions between aroids and arboreal mammals in the brazilian Atlantic rainforest. Plant Ecology 145:75-82.

Vieira, E. M., M. A. Pizo, and P. Izar. 2003. Fruit and seed exploitation by small rodents of the brazilian Atlantic forest. Mammalia 67:533-539.

GENERAL METHODS

FOCUS SPECIES

In the analysys I concentrated on the rodent species *Akodon montensis* (Grass mouse, Thomas 1902), *Oligoryzomys nigripes* (Black-footed pigmy rice rat, Olfers 1818), and *Delomys sublineatus* (Pallid Atlantic forest rat, Thomas 1903). In the investigation of movement distances (Chapter 1) I also included *Oryzomys russatus* (Rice rat, Wagner 1848) and *Thaptomys nirita* (Blackish grass mouse, Lichtenstein 1829). Marsupials species investigated were the mouse opossums *Marmososps incanus* (Grey slender mouse opossum, Lund 1840) and *Gacilinanus microtarsus* (Brazilian gracile mouse opossum, Wagner 1842; see Fig 5). The rodents *D. sublineatus*, *T. nirita*, and both marsupials are endemic to the Atlantic Forest while the other species occur in other biomes as well. As indicated by the previous investigations on community structure by Dr. Renata Pardni and her students *A. montensis*, *O. nigripes*, and *G. microtarsus* (generalist species) seemed to be less affected by fragmentation compared to *D. sublineatus*, *M. incanus*, *O. russatus*, and *T. nigrita* (specialist species).

STUDY AREA

The study area is located about 40 km southeast from the City of São Paulo, Brazil, in the municipalities of Cotia and Ibiúna. It comprises a continuous forest of about 10.000 km^2 (Morro Grande Reserve) connected to a large continuous forest area in the Serra do Mar and an adjacent fragmented landscape (Fig. 3). Altitude varies from 850 to 1100 m above sea level (Oliveira-Filho and Fontes 2000) and the temperature ranges from 27° C (mean maximum) to 11° C (mean minimum). Rainfall is around 1300–1400 mm/year with the driest and coldest months between April and August. All forest fragments and the continuous forest (except some small regions) are about 80 years old. Within this landscape about 30 % is covered with forest and the fragments range from 4 to 270 ha in size. Some are connected by corridors to other fragments while others are isolated by agricultural fields. The matrix between the remaining forest fragments contains mainly agricultural fields for vegetable cultivation (33 % of total landscape). Further, plantations of pine and eucalyptus cover about 7 % of the landscape as well as native vegetation in early stages of regeneration. The area was chosen because of its homogeneity in terms of climate, forest type and forest age and the opportunity to investigate the effects of fragmentation in fragments with varying degree of size, connectivity, edges and different surrounding matrix in comparison to a control area of almost undisturbed forest.

Due to limited time and resources, I concentrated in my study on five forest fragments and the Morro Grande Reserve (Fig. 3). The fragments differed in size (two small fragments about 15 ha, two medium sized fragments about 30 ha and one large fragment of 175 ha), shape, connectivity and proportion of human-altered area surrounding the fragment. In the Morro Grande reserve a control area in

secondary forest was chosen. Thereby, I was able to ensure that probable differences in small mammal response to fragmentation were not caused by differences in forest age.

SMALL MAMMAL TRAPPING

To investigate small mammal populations as well as ecological

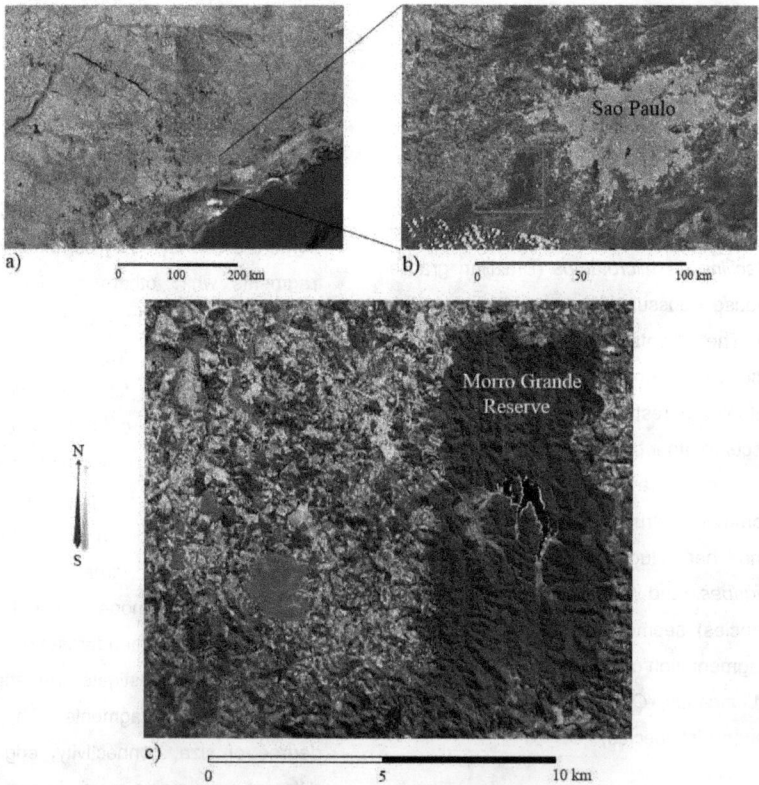

Fig. 3 Study area. a) state of Sao Paulo b) Sao Paulo City with study area marked in the red square c) Study area with the Morro Grande Reserve and adjacent fragmented landscape. Red areas mark the fragments within which small mammals where captured and the study site inside the Reserve, respectively.

Fig. 4 Sherman life traps positioned on a tree branch (upper picture) and on the ground (lower picture)

parameters such as movement patterns and use of vegetation I established a grid system of 10 x 10 trap locations in each of the chosen fragments. Each grid covered an area of approximately 3.2 ha within which 200 live traps (Sherman Traps Inc., Tallahassee, USA; Fig. 4) were placed in equal distances to each other (20 m). Capturing animals with live traps has several advantages. First of all, when compared to other methods, this method is relatively gentle for the captured animals. The individuals are protected from weather and predators inside the trap. Secondly,

by using Sherman live traps, I was able to collect faeces (for parasitological investigations, Chapter 4) out of the traps and since only one individual per trap was caught, the correct assignment to the animal was possible. This is not the case in other types of traps where several individuals are caught in the same trap and correct assignment of faeces to individual is impractical. Thirdly, capturing with Sherman traps is selective for small mammals and very few species are captured which do not belong to this group (only four birds and some toads during this study).

By using a grid system for trap location I was able to identify mean distances moved for individuals captured more than once. Although these mean distances are broad scaled due to the fixed distance of traps, one can use them as an index for movement distances (Chapter 1). By relating the number of captured animals to the area covered by the grid it is possible to calculate an estimate of species density for the different fragments (Chapter 2). Further, the grid comprised different structural features and vegetation within each study site. Some species might be caught preferentially in areas with certain vegetation structure, while another species seems to avoid them. With additional recording of vegetation parameters in each of the 600 trap locations I was able to identify preferred vegetation features of different species (Chapter 3).

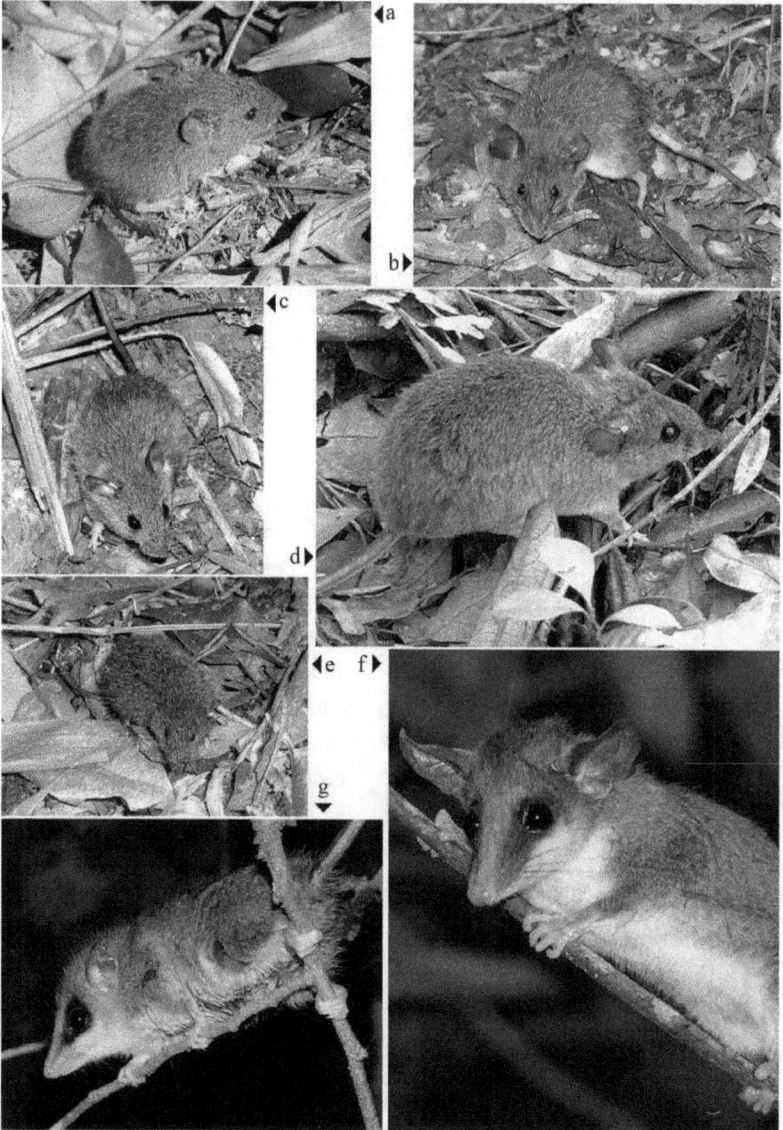

Fig 5. Investigated species: a) *Akodon montensis*; b) *Delomys sublineatus*; c) *Oligoryzomys nigripes*; d)*Oryzomys russatus*; e)*Thaptomys nigrita*; f) *Marmosops incanus*; g)*Gracilinanus microtarsus*

REFERENCES

Oliveira-Filho, A. T., and M. A. L. Fontes. 2000. Patterns of floristic differentiation among Atlantic Forests in Southeastern Brazil and the influence of climate. Biotropica 32:793-810.

CHAPTER 1

Movement distances of five rodent and two marsupial species in forest fragments of the coastal Atlantic forest

ABSTRACT

Movement distances provide information on diverse population biological parameters and are essential in understanding the ecology of a species. Mean distances moved between successive captures (SD), distribution of movement distances, and the mean maximum distances moved (MMDM) were investigated in five rodent and two marsupial species in forest fragments of the coastal Atlantic forest (Mata Atlântica) in the state of São Paulo, Brazil. The investigated species were the rodents *Akodon montensis* (Thomas 1902), *Oligoryzomys nigripes* (Olfers 1818), *Delomys sublineatus* (Thomas 1903), *Oryzomys russatus* (Wagner 1848), and *Thaptomys nigrita* (Lichtenstein 1829), and the marsupials *Marmosops incanus* (Lund 1840), and *Gracilinanus microtarsus* (Wagner 1842). *Akodon montensis* differed significantly from all other species and moved the lowest SD and MMDM. The marsupials differed significantly from most of the rodents and moved the largest SDs. All species showed the highest frequency of movements in a distance class of 0-20 m. Differences between sexes in SD was detected for *O. nigripes*, *O. russatus*, and *G. microtarsus*, males moving significantly longer distances than females. The different study sites had no influence on SD in any of the investigated species. Only the males of *G. microtarsus* showed a seasonal variation in SD, moving longer distances during reproductive activity.

INTRODUCTION

Movements of animals are an important element in their ecology (Turchin 1991, Diffendorfer *et al.* 1995, Slade & Russell 1998) and provide essential information about the spatial distribution of species (Stapp & Van Horne 1997). Movement patterns and distances are related to several aspects of the ecology of a species, like genetic structure, feeding habits, food availability, mating systems, and reduction to exposure to predators (Slade & Swihart 1983, Austad & Sunquist 1986, Sunquist *et al.* 1987, Barnum *et al.* 1992, Stapp & Van Horne 1997, Roche *et al.* 1999) Moreover, movements are associated with the social organization and population dynamics of a species (Slade & Swihart 1983, Bowers *et al.* 1996, Mendel & Vieira 2003, Bergallo & Magnusson 2004). The knowledge of distances moved by different species is necessary to calculate basic parameters of the ecology of a population, such as density estimates, as well as to understand its genetic structure (Mendel & Vieira 2003).

The use of distances moved between successive captures instead of area calculations like minimum convex polygons has several advantages. Movement distances are comparatively easy to assess, easy to calculate, and can be used even for individuals with capture histories too brief for models (Stickel 1954, Slade & Swihart 1983). Studies have indicated that distance measurements are correlated with home range size (Faust et al. 1971, Slade & Russell 1998). They provide an average index of the home range for the species (Davis 1953) and can be used for comparison between groups within a species (e.g., age- or sex-specific analyses, Slade & Russell 1998) or between different species (Gentile & Cerqueira 1995). Nevertheless, using trapping distances as an index for spatial use also has some disadvantages. Some individuals might preferentially be caught in one trap or make a trap unattractive for others. The capture probabilities between individuals or species might differ because of different behavioral responses or heterogeneity due to age or sex, or to time effects (White et al. 1982). Therefore, the use of distances between successive captures to reflect movements of species is not directly comparable to home range size data obtained by direct measurements (e.g., radio-tracking).

The coastal Atlantic forest of Brazil (Mata Atlântica) is one of the most diverse but at the same time one of the most threatened environments in the world (Myers et al. 2000, Galindo-Leal & de Gusmão Câmara 2003). Only about 8 % of its original extent still remains and it harbors a great variety of small mammal species (Emmons & Feer 1997, Stevens & Husband 1998) with an overall high percentage of endemic species (Fonseca et al. 1996, Stevens & Husband 1998). For most Mata Atlântica species information about distances moved is limited.

During a capture-recapture study, in the course of a study on population dynamics of small mammals inhabiting the coastal Atlantic forest, I investigated the distances moved between successive captures (SD, Murie & Murie 1931), the distribution of movement distances, and the mean maximum distance moved (MMDM, Wilson & Anderson 1985) of seven small mammal species in a trapping grid. I focussed on the differences in movement length (SD, MMDM) between species and further investigated differences in SD between gender, study sites, and capture sessions.

METHODS

STUDY AREA

The study was conducted in the region of Caucaia do Alto (23°40′ S, 47°01′ W), situated in the municipalities of Cotia and Ibiúna, São Paulo state, about 80 km south-west of the city of São Paulo, Brazil, in a transition zone between dense ombrophilous forest and semi-deciduous forest classified as "Lower Montane

Atlantic rainforest" (Oliveira-Filho & Fontes 2000). The altitude varies between 800 and 1100 m (Ross & Moroz 1997). Monthly mean temperature ranges from a minimum of 11°C to a maximum of 27°C. Annual precipitation equals 1300-1400 mm and fluctuates seasonally with the driest and coldest months between April and August.

The area includes a fragmented area and a large, lower mountainous Atlantic forest area (Morro Grande Reserve). The fragmented area consists of secondary forest fragments embedded in an agricultural landscape. Secondary forest covers 31 % of the landscape, which is dominated by anthropogenic habitat (agricultural fields: 33 %; areas with rural buildings or urban areas: 15 %; vegetation in early stages of regeneration: 10 %; pine and eucalyptus plantations: 7 %; others: 4%). The Morro Grande Reserve also consists mainly of secondary forest. Only a minor part of the reserve provides mature forest.

Animals were captured in five forest fragments of different sizes (14-175 ha), and in the Morro Grande Reserve (10 700 ha). All sites are of secondary growth forest and between 50 and 80 years of age (Godoy Teixeira 2005). Two fragments are about 14 ha (S1, S2), another two fragments are approximately twice this size (30 ha; S3, S4), and the fifth fragment (S5) is the largest, embracing 175 ha of secondary forest. The control site (CS) also consists of secondary forest and is part of the Morro Grande Reserve. A more detailed description of the study area can be found in Pardini et al. (2005).

DATA COLLECTION

A regular trapping grid of one hundred trap stations 20 m apart was established in all six study sites. The grid consisted of 10 x 10 trap stations, except in two sites where the shape of the fragments and the general objective of placing all traps within the forest required that the trapping grid be slightly modified. Two live-traps (H. B. Sherman Traps Inc., Tallahassee, USA) were set at each trap station, one small (23 x 9 x 8 cm) and one large (38 x 11 x 10 cm). One of the traps was set on the ground, and the other at about 1.0 to 1.5 m above the ground, alternating the positions of small and large traps between trap stations.

Data were collected during five trapping sessions in each of the study sites from July 2003 to March 2005 (1st session: 23.07.-19.09.2003, 2nd session: 27.09.- 7.11.2003, 3rd session: 4.03.-7.04.2004, 4th session: 18.05.-26.06.2004, 5th session: 27.01.-3.03.2005). Each trapping session consisted of six nights of capture, for a total of 6000 trap-nights per study site. Traps were baited with banana and a mixture of peanut butter, oat, and sardines, left open for the night, checked every morning and rebaited if necessary. Captured animals were anesthetized (Forene®, Abbott GmbH, Wiesbaden, Germany) for 1-2 minutesand marked individually by numbered ear tags (Fish and small animal tag size 1, National Band and Tag Co., Newport, Kentucky, USA). In addition to sexing and weighing, the length of their tibia was measured to the nearest 0.5 mm. All individuals were released subsequently at their respective trapping location.

STATISTICAL ANALYSIS

All individuals captured more than once were included in the calculations of SD. The SD was compared between species, sexes, study sites, and between sessions (seasonal variation). To investigate the distribution of movement distances, the distances were grouped into classes. The maximum distance moved was calculated for each individual at each capture session, and the mean maximum distance moved (MMDM) was calculated for each species.

Non-parametric statistics were used because distances were not normally distributed. To account for multiple comparisons Bonferroni correction was applied (Sachs 1992). All tests were conducted on SPSS 11.5.1 (SPSS Inc., Chicago, USA) using a significance level of 0.05.

RESULTS

SPECIES CAPTURED

A total of 1338 individuals belonging to 13 species were captured 3031 times (Table 1), leading to a mean trapping success of 8.4 % in 36 000 trap-nights. Movement distances were calculated for the rodents *Akodon montensis* (Thomas 1902), *Oligoryzomys nigripes* (Olfers 1818), *Delomys sublineatus* (Thomas 1903), *Oryzomys russatus* (Wagner 1848) and *Thaptomys nigrita* (Lichtenstein 1829), as well as for the marsupials *Marmosops incanus* (Lund 1840) and *Gracilinanus microtarsus* (Wagner 1842). Only *A. montensis* and *M. incanus* were captured at all six study sites and in every capture session. The didelphid marsupials *Didelphis aurita* (Wied-Neuwied 1826), *Monodelphis americana* (Müller 1776), and *Micoureus paraguayanus* (Oken 1816), and the rodents *Juliomys pictipes* (Osgood 1933), *Oryzomys angouya* (Fischer 1814), and *Brucepattersonius* aff. *iheringi* (Thomas 1896) were only occasionally captured and therefore excluded from this analysis (Table 1).

DIFFERENCES BETWEEN SPECIES IN MOVEMENT PATTERNS

The seven species differed significantly in their distance moved between successive captures (SD, Table 2; Kruskal-Wallis-Test, χ^2 = 158.07; df = 6; p < 0.0001). Pairwise comparison (Mann-

Tab. 1 Taxonomy, body mass, locomotion habits, and numbers of individuals captured of each species (males, females, and unsexed individuals in parenthesis). Taxonomy, body mass, and locomotion habits are based on Musser & Carleton (1993), Tyndale-Biscoe (2005), and personal observations.

Species	Family	Subfamily	Body mass	Locomotion	Number of individuals captured
Akodon montensis	Muridae	Sigmodontinae	19-57 g	terrestrial	476 (252/213/11)
Thaptomys nigrita	Muridae	Sigmodontinae	12-30 g	fossorial	79 (44/33/2)
Oligoryzomys nigripes	Muridae	Sigmodontinae	9-40 g	scansorial	158 (99/55/4)
Delomys sublineatus	Muridae	Sigmodontinae	20-75 g	terrestrial	153 (75/73/5)
Oryzomys russatus	Muridae	Sigmodontinae	40-120 g	terrestrial	43 (24/17/2)
Marmosops incanus	Didelphidae	Thylaminae	13-140 g	semi-arboreal	144(72/69/3)
Gracilinanus microtarsus	Didelphidae	Thylaminae	19-29 g	arboreal	140 (75/59/6)
Didelphis aurita	Didelphidae	Didelphinae	700-1500 g	scansorial	42 (15/24/3)
Monodelphis americana	Didelphidae	Marmosinae	23-35 g	terrestrial	14 (8/6/0)
Micoureus paraguayanus	Didelphidae	Marmosinae	80-150 g	scansorial	2 (0/2/0)
Oryzomys angouya	Muridae	Sigmodontinae	40-120 g	scansorial	35 (17/14/4)
Juliomys pictipes	Muridae	Sigmodontinae	11-30 g	arboreal	32 (18/11/3)
Brucepattersonius aff. iheringi	Muridae	Sigmodontinae	11-35 g	terrestrial/ fossorial	20 (13/6/0)
Total					1338

Whitney U-Test) showed differences between the movements of *A. montensis* and all other species (all Z > -3.07; all p < 0.002), between *T. nigrita* and *M. incanus* (Z = - 2.78; p = 0.005), *T. nigrita* and *G. microtarsus* (Z = -3.19; p < 0.001), *O. nigripes* and *M. incanus* (Z = -4.21; p < 0.0001), *O. nigripes* and *G. microtarsus*. (Z = -4.43; p < 0.0001), *D. sublineatus* and *M. incanus* (Z = - 2.72; p = 0.007), and *D. sublineatus* and *G. microtarsus* (Z = -3.56; p < 0.0001). The former species always moved smaller distances than the latter (e.g., *T. nigrita* moved smaller distances than *M. incanus*; all Bonferroni adjusted α = 0.008). The MMDM (Table 2) was smallest for *A. montensis* and differed significantly from other species (all Z > - 2.66; all p < 0.008) except *O. nigripes* (Z = -1.68; p = 0.093). The two marsupial species *M. incanus* and *G. microtarsus* moved significantly greater MMDM than the rodent species (all Z > -2.95; all p < 0.0001) except *O. russatus* (Z = -0.90; p = 0.368 and Z = -2.13; p = 0.034, all Bonferroni adjusted α = 0.008).

Tab. 2 Number of individuals recaptured, number of captures, mean distance moved between successive captures (SD), and mean maximum distance moved (MMDM), in meters, with standard errors for the different species. In case of significant sexual differences the sexes are separated.

Species	Number of individuals recaptured	Total number of captures (including recaptures)	Mean distance moved between successive captures (SD)	Mean maximum distance moved (MMDM)
A. montensis	392	685	19.54 ± 1.05	26.30 ± 1.37
T. nigrita	28	41	29.98 ± 3.93	38.29 ± 5.65
O. nigripes				
males	46	55	31.81 ± 3.67	32.95 ± 3.96
females	16	26	18.04 ± 4.64	23.39 ± 4.85
D. sublineatus	83	118	32.42 ± 2.33	38.14 ± 3.12
O.russatus				
males	19	30	39.71 ± 5.18	51.71 ± 7.25
females	12	28	23.15 ± 6.13	53.51 ± 19.23
M. incanus	88	158	41.07 ± 2.33	53.75 ± 3.15
G. microtarsus				
males	35	61	65.22 ± 7.54	84.60 ± 8.67
females	30	68	38.86 ± 4.46	59.89 ± 7.67

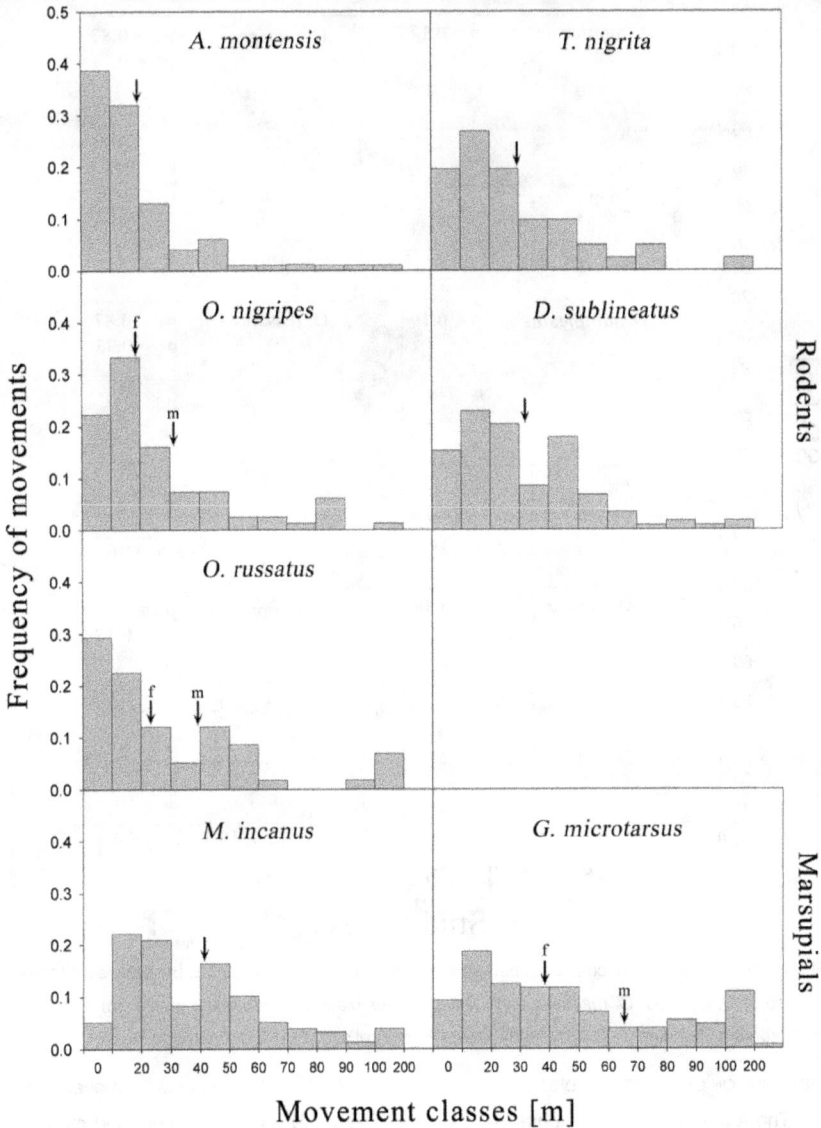

Fig. 6 Distribution of movement distances (in meters) between successive captures for five rodents and two marsupials of the coastal Atlantic rainforest. The arrows mark the mean distance moved between successive captures for the species (m = males, f = females in case of significant difference between sexes). Rodents: *Akodon montensis*, *Oligoryzomys nigripes*, *Delomys sublineatus*, *Oryzomys russatus*, *Thaptomys nigrita*; marsupials: *Marmosops incanus*, *Gracilinanus microtarsus*.

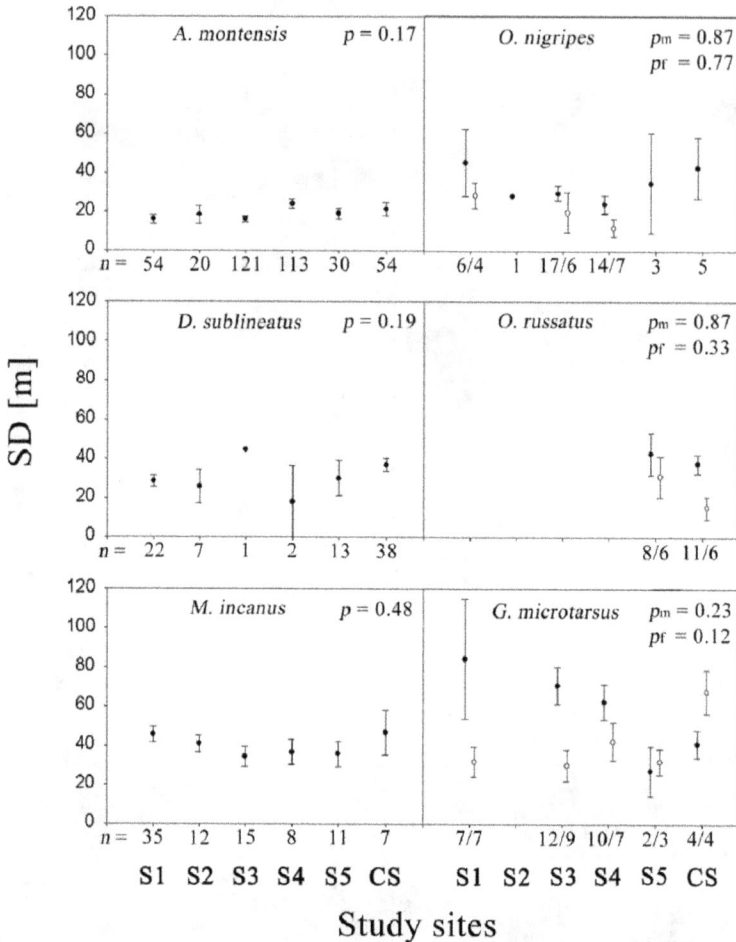

Fig. 7 Differences in distance moved between successive captures (SD) ± SE between study sites. In case of *O. nigripes*, *O. russatus*, and *G. microtarsus* black dots represent males (m) and white dots represent females (f). S1-S5: study sites, CS: control site in the Morro Grande Reserve.

DISTRIBUTION OF MOVEMENT DISTANCES

The rodent species *A. montensis* and *O. russatus* were most frequently recaptured at the same station as first capture (Fig. 6). All other species showed the highest frequency of movements in the distance class of 20 m.

All rodent species moved most frequently short distances, and more than 80% of their movements did not exceed 50 m. The distribution of movements was less concentrated for the marsupials *M. incanus* and *G. microtarsus*. Both were rarely recaptured at the same trap station and more than 27 % of all movements exceeded 50 m in *M. incanus*. In *G. microtarsus*, almost 50 % of the movements were longer than 50 m. It moved long distances more frequently than the other species, and the longest movement of all, over 200 m, was recorded for this species (Fig 1).

DIFFERENCES BETWEEN SEXES, STUDY SITES, AND SESSIONS

Only the movements of *O. nigripes*, *O. russatus* and *G. microtarsus* differed significantly between males and females (*O. nigripes*: $Z = -2.04$; $p = 0.042$; *O. russatus*: $Z = -2.06$; $p = 0.039$; *G. microtarsus*: $Z = -2.79$; $p = 0.005$). For these species SD was smaller for females than for males (Table 2). The six study sites did not differ significantly in SD for any of the investigated species (all $x^2 < 7.76$; $df = 1-5$ because not all species were captured in all study sites; all $p > 0.05$, Fig. 7). For *T. nigrita* this test was not performed because this species only occurred in the Morro Grande Reserve.

A. montensis, *T. nigrita*, *D. sublineatus* and *M. incanus* did not differ significantly between capture sessions in SD. For *O. nigripes*, *O. russatus* and *G. microtarsus*, the test was performed separately for the two sexes because of differences in mean distance moved. Both sexes of *O. nigripes* and *O. russatus* showed no significant difference between capture sessions. Males of *G. microtarsus* could only be

compared between three of the capture sessions due to insufficient captures in the other capture sessions ($x^2 = 9.20$; $df = 2$; $p = 0.01$). They moved significantly longer distances in the fourth capture session ($SD_4 = 80.57$ m, 18.05. - 26.06.2004, beginning of dry season) than in the second ($SD_2 = 43.89$ m, 27.09. - 7.11.2003, beginning of wet season; $Z = -2.66$; $p = 0.007$, Bonferroni adjusted $\alpha = 0.025$). For females, the number of individuals captured more than once in two of the study sites was too small for statistical analysis.

DISCUSSION

Distinct movement patterns were revealed for the species studied. Hypothetically, the high abundance of *A. montensis* might indicate high resource availability for that species in the study area, which in turn could be a reason for low distances moved, although this can only be tested with additional data on resource availability in the study area. However, other investigations on closely related species of the genus *Akodon* revealed similar distances moved, such as *A. curso*r in a coastal shrubland in the state of Rio de Janeiro (highest frequency of movements between 0 and 20 m, more than 50% of movements less than 30 m, Gentile & Cerqueira 1995), in Atlantic forest fragments (highest frequency of movements between 0 and 20 m (Pires *et al.* 2002), and in a gallery forest in the *cerrado* of central Brazil (mean distance moved: 32.4 ± 37.2 m, Nitikman & Mares 1987), and *A. montensis* in secondary forest in Rio de Janeiro (over 70 % of captures between 0 and 40 m; mean distance moved : 40.1 ± 53.5 m, Davis 1945). Our study does not confirm the general pattern of sexual dimorphism in home ranges in akodontines, males having larger home ranges (review in Gentile *et al.* 1997). I did not find any significant differences in movement patterns between sexes for *A. montensis*.

T. *nigrita* is the smallest of the investigated species. It is fossorial and vanished almost always underneath foliage immediately after release. Considering these habits, SD of about 30 m (Table 2) is unexpectedly high. However more than 40 % of all movements did not exceed 20 m.

The SD and the distribution of movements of *O. nigripes* are similar to the results of other studies for this species (Nitikman & Mares 1987: 27.3 ± 22.7 m; Pires *et al.* 2002: movements did not exceed 80 m). Males moved larger mean distances than females. To our knowledge, this is the first evidence of sex differences in movement distances in *O. nigripes*. One can speculate that this difference might indicate a promiscuous mating system, with larger male home ranges including more than one female

home range, although direct home range data are needed to test this hypothesis.

Davis (1945) captured *D. sublineatus* usually close to the original capture site, with the majority of captures being less than 50 m. This coincides with our results. Over 80% of all movements of *D. sublineatus* were less than 50 m.

Oryzomys russatus showed differences in SD between the two sexes. The values of the SD were significantly smaller for females than for males. In other studies in two different study areas in the Atlantic forest of São Paulo state, no difference was found between home ranges (convex polygon method) of males and females of *O. russatus* (Bergallo 1995, Bergallo & Magnusson 2004), hypothesizing a monogamous mating system in *O. russatus*. In our study, the capture points of the different female individuals were at least 50 m apart when more than one female was captured during one capture session, implying that there is little overlap at the areas used by different females. In some cases different males were captured close to or in the same capture point of females. These results support the findings of Bergallo (1995), showing little intrasexual but much intersexual spatial overlap, although more intense data collection is needed for further investigation of the mating system of *O. russatus* in the study area.

The pattern of larger distances moved by marsupials was confirmed in this study

(Nitikman & Mares 1987, Fonseca & Kierulff 1989, Pires *et al*. 2002). Fonseca & Kierulff (1989) captured mostly males of *M. incanus* and found a slightly higher SD of 64.7 m compared with our results (41.07 m). I found no significant differences in SD between sexes for *M. incanus*, but did for *G. microtarsus*. The MMDM was also significantly higher for males than for females of this species. This might suggest a sexual dimorphism in movement distances of *G. microtarsus*, probably with larger male home ranges. Larger male home ranges in marsupials in Brazil were recorded for *Micoureus demerarae* (Pires *et al*. 1999). No differences in movements between successive captures could be determined in *Didelphis aurita*, *Philander frenatus*, and *Metachirus nudicaudatus* by Gentile & Cerqueira (1995). On the other hand, Loretto & Vieira (2005) detected seasonal differences between sexes in movements of *Didelphis aurita*. Compared to the mean distance moved by the ecologically similar species *Marmosa agilis* recorded by Nitikman & Mares (1987) in *cerrado* gallery forest (41.1 ± 35.0 m), *G. microtarsus* traveled similar mean distances in our study. *G. microtarsus* was the only species that showed temporal differences in SD. Males moved larger distances at the beginning of the dry season 2004 than at the beginning of the rainy season 2003. The interpretation of this result is difficult. *G. microtarsus* has

an insectivorous-omnivorous diet (Martins and Bonato 2004). Further investigation on food availability in the study area is needed to evaluate a possible influence of food abundance on the observed movement distances. In general, arthropod and fruit abundance in the rainy season is higher compared with the dry season (Janzen 1973, Develey & Peres 2000, R. Pardini, pers. comm.). Hence movement distances should be smaller in the rainy season to assure a sufficient food supply. More likely is an influence of reproductive activity. Seasonal reproduction in the rainy season has been observed for several marsupials in the Neotropics (Fonseca & Kierulff 1989, Bergallo 1994). The males of *G. microtarsus* might travel greater distances in the rainy season in search of females.

I found no evidence that fragment size influences the movement distances of the species. None of the investigated species showed a significant difference between study sites.

The results of this study can help us understand the population ecology of small mammal species in the coastal Atlantic forest of Brazil. Additional information will be gained by further investigations into the relationship between population size and the respective movement distance of the different species.

REFERENCES

Austad, S.N., & M.E. Sunquist. 1986. Sex-ratio manipulation in the common opossum. Nature 324: 58-60.

Barnum, S.A., Manville, C.J., Tester, J.R., & W.J. Carmen. 1992. Path selection by Peromyscus leucopus in the presence and absence of vegetative cover. Journal of Mammalogy 73: 797-801.

Bergallo, H. 1994. Ecology of a small mammal community in an Atlantic rainforest area in southeastern Brazil. Studies on Neotropical Fauna and Environment 29: 197-217.

Bergallo, H. 1995. Comparative life-history characteristics of two species of rats, Proechimys iheringi and Oryzomys intermedius, in an Atlantic forest of Brazil. Mammalia 59: 51-64.

Bergallo, H.G., & W. E. Magnusson. 2004. Factors affecting the use of space by two rodent species in Brazilian Atlantic forest. Mammalia 68: 121-132.

Bowers, M.A., Gregario, K., Brame, C.J., Matter, S.F., & J.L. Dooley Jr. 1996. Use of space and habitats by meadow voles at the home range, patch and landscape scales. Oecologia 105: 107-115.

Davis, D.E. 1945. Home ranges of some Brazilian mammals. Journal of Mammalogy 26: 119-127.

Davis, D.E. 1953. Analysis of home range from recapture data. Journal of Mammalogy 34: 352-358.

Develey, P.F., & C.A. Peres. 2000. Resource seasonality and the structure of mixed species bird flocks in a coastal Atlantic forest of southeastern Brazil. Journal of Tropical Ecology 16: 33-53.

Diffendorfer, J.E., Gaines, M.S., & R.D. Holt. 1995. Habitat fragmentation and movements of three small mammals (Sigmodon, Microtus, and Peromyscus). Ecology 76: 827-839.

Emmons, L.H., & F. Feer. 1997. Neotropical rainforest mammals: a field guide, 2nd edition. The University of Chicago Press, Chicago.

Faust, B.F., Smith, M.H., & W.B. Wray. 1971. Distances moved by small mammals as an apparent function of grid size. Acta Theriologica 16: 161-177.

Fonseca, G.A.B., Herrmann, G., Leite, Y.L.R., Mittermeier, R.A., Rylands, A.B., & J.L. Patton. 1996. Lista anotada dos mamíferos do Brasil. Occasional Papers in Conservation Biology 4: 1-38.

Fonseca, G.A.B., & M.C.M. Kierulff. 1989. Biology and natural history of Brazilian Atlantic forest small mammals. Bulletin of the Florida State Museum, Biological Sciences 34: 99-152.

Galindo-Leal, C., & I. de Gusmão Câmara. 2003. The Atlantic forest of South America: biodiversity status, threats, and outlook. Island Press, Washington.

Gentile, R., & R. Cerqueira. 1995. Movement patterns of five species of small mammals in a Brazilian restinga. Journal of Tropical Ecology 11: 671-677.

Gentile, R., D'Andrea, P.S., & R. Cerqueira. 1997. Home ranges of Philander frenata and Akodon cursor in a Brazilian restinga (coastal shrubland). Mastozoologia Neotropical 4: 105-112.

Godoy Teixeira, A.M. 2005. Análise da dinâmica da paisagem e de processos de fragmentação e regeneração na região de Caucaia-do-Alto, SP (1962-2000). University of São Paulo, São Paulo.

Janzen, D.H. 1973. Sweep samples of tropical foliage insects: effects of seasons, vegetation types, elevation, time of day and insularity. Ecology 54: 687-708.

Loretto, D., & M.V. Vieira. 2005. The effects of reproductive and climatic seasons on movements in the black-eared opossum (Didelphis *aurita* Wied-Neuwied, 1826). Journal of Mammalogy 86: 287-293.

Martins, E.G., & V. Bonato. 2004. On the diet of *Gracilinanus microtarsus* (Marsupialia, Didelphidae) in an Atlantic rainforest fragment in southeastern Brazil. Mammalian Biology 69: 58-60.

Mendel, S.M., & M.V. Vieira. 2003. Movement distances and density estimation of small mammals using the spool-and-line technique. Acta Theriologica 48: 289-300.

Murie, O.J., & A. Murie. 1931. Travels of *Peromyscus*. Journal of Mammalogy 12: 200-209.

Musser, G.G. & M.D. Carleton, 1993. Family Muridae. Pp. 501-753 *in* Wilson, D.E. & D.M. Reeder (eds.). Mammal species of the world: a taxonomic and geographic reference, second edition. Smithsonian Institution Press, Washington and London.

Myers, N., Mittermeier, R.A., Mittermeier, C.G., Fonseca, G.A.B., & J. Kent. 2000. Biodiversity hotspots for conservation priorities. Nature 403: 853-858.

Nitikman, L.Z., & M.A. Mares. 1987. Ecology of small mammals in a gallery forest of Central Brazil. Annals of Carnegie Museum 56: 75-95.

Oliveira-Filho, A.T., & M.A.L. Fontes. 2000. Patterns of floristic differentiation among Atlantic forests in southeastern Brazil and the influence of climate. Biotropica 32: 793-810.

Pardini, R., Marques de Souza, S., Braga-Neto, R., & J.P. Metzger. 2005. The role of forest structure, fragment size and corridors in maintaining small mammal abundance and diversity in an Atlantic forest landscape. Biological Conservation 124: 253-266.

Pires, A.S., Fernandez, F.A.S., & D. de Freitas. 1999. Patterns of space use by *Micoureus demerarae* (Marsupialia: Didelphidae) in a fragment of Atlantic forest in Brazil. Mastozoologia Neotropical 6: 39-45.

Pires, A.S., Koeler Lira, P., Fernandez, F.A.S., Schittini, G.M., & L.C. Oliveira. 2002. Frequency of movements of small mammals among Atlantic coastal forest fragments in Brazil. Biological Conservation 108: 229-237.

Roche, B.E., Schulte-Hostedde, A.I., & R.J. Brooks. 1999. Route choice by deer mice (*Peromyscus maniculatus*): reducing the risk of auditory detection by predators. American Midland Naturalist 142: 194-197.

Ross, J.L.S., & I.C. Moroz. 1997. Mapa Geomorfológico do Estado de São Paulo: escala 1:500.000. FFLCH-USP. IPT and Fapesp, São Paulo.

Sachs, L. 1992. Angewandte Statistik. Springer Verlag, Berlin.

Slade, N.A., & L.A. Russell. 1998. Distances as indices to movements and home-range size from trapping records of small mammals. Journal of Mammalogy 79: 346-351.

Slade, N.A., & R.K. Swihart. 1983. Home range indices for the hispid cotton rat (*Sigmodon hispidus*) in Northeastern Kansas. Journal of Mammalogy 64: 580-590.

Stapp, P., & B. Van Horne. 1997. Response of Deer Mice (*Peromyscus maniculatus*) to shrubs in shortgrass prairie: linking small-scale movements and the spatial distribution of individuals. Functional Ecology 11: 644-651.

Stevens, S.M., & T.P. Husband. 1998. The influence of edge on small mammals: evidence from Brazilian Atlantic forest fragments. Biological Conservation 85: 1-8.

Stickel, L.F. 1954. A comparison of certain methods of measuring ranges of small mammals. Journal of Mammalogy 35: 1-15.

Sunquist, M.E., Austad, S.N., & F. Sunquist. 1987. Movement patterns and home range in the common Opossum (*Didelphis*

marsupialis).   Journal   of
Mammalogy 68: 173-176.

Turchin, P. 1991. Translating foraging movements in heterogeneous environments into spatial distribution of foragers. Ecology 72: 1253-1266.

Tyndale-Biscoe, H., 2005. Life of Marsupials. CSIRO Publishing, Collingwood.

White, G.C., Anderson, D.R., Burnham, K.P., & D.L. Otis. 1982. Capture-recapture and removal methods for sampling closed populations. Los Alamos National Laboratory, Los Alamos.

Wilson, K.R., & D.R. Anderson. 1985. Evaluation of two density estimators of small mammal population size. Journal of Mammalogy 66: 13-21.

48

CHAPTER 2

Fragmentation effects on population density of three rodent and two marsupial species in secondary Atlantic forest

ABSTRACT

I investigated the population density of the two common rodent species *Akodon montensis*, *Oligoryzomys nigripes*, the rare, more specialized endemic *Delomys sublineatus* and the two endemic marsupial species *Marmosops incanus* (specialist) and *Gracilinanus microtarsus* (generalist) in secondary forest fragments of the coastal Atlantic forest. Linear regression was used to examine possible relationships between population density and four landscape variables. I tested the hypothesis that specialist species which are more effected by fragmentation effects than the common species should decrease in population density with decreasing fragment size while the density of common generalist species should be unaffected or even increase in smaller forest remnants. The results revealed distinct differences between the investigated species. While *A. montensis* as well as *O. nigripes* showed no significant association to any landscape variable, the density of *D. sublineatus* was associated positively to fragment size as well as negatively to edge density. Results on marsupials were less clear. The population density of the more specialized *M. incanus* was unexpectedly negatively correlated to the degree of connectivity, indicating some other aspects influencing density of this species. Density of *G. microtarsus* did not show any correlation to landscape variables. The results show the importance of large and connected forest remnants in the fragmented coastal Atlantic forest to guarantee an effective protection of rare, endemic small mammal species like *D. sublineatus*.

INTRODUCTION

The coastal Atlantic forest (Mata Atlântica) in Brazil is one of the most diverse and most threatened biomes in the world (Myers et al. 2000). Due to severe human impact over the last few centuries most of the primary coastal Atlantic forest has been destroyed. Only 8% of its original extent yet remains and the remaining forest is highly fragmented (Terborgh 1992, Galindo-Leal and de Gusmão Câmara 2003, Tabarelli et al. 2005). This does not only include primary forest, but also secondary forest remnants in different stages of regeneration (Galindo-Leal and de Gusmão Câmara 2003). The Mata Atlântica harbours a large number of endemic species. 160 birds, 128 amphibians and 73 mammals are restricted to this ecosystem (Ayres et al. 2005). 92 species of small mammals are known to inhabit the Mata Atlântica, of which 43 are endemic (Fonseca et al. 1996). Small mammals play a key role in forest ecosystems in terms of seed dispersal (Forget 1991, Sánchez-Cordero and Martínez-Gallardo 1998, Brewer and Reimánek 1999, Vieira and Izar 1999,

Vieira et al. 2003, Pimentel and Tabarelli 2004) and the dispersal of mycorrhizal fungi (Janos and Sahley 1995). Furthermore, they affect the forest's regeneration by predation on seedlings (Forget 1993, Pizo 1997, Sánchez-Cordero and Martínez-Gallardo 1998, Pires and Fernandez 1999, Forget et al. 2000, Pires et al. 2002, Vieira et al. 2003). Small mammals are impaired by fragmentation with varying intensity. Whereas some species decline in abundance in more isolated fragments others might not be influenced or even benefit from the consequences of fragmentation (Laurance 1991, Pardini 2004, Viveiros de Castro and Fernandez 2004, Pardini et al. 2005, Umetsu and Pardini *in press*). A decline in abundance and decreasing population density in small and isolated habitat patches might lead to demographic stochasticity, environmental stochasticity, edge effects and inbreeding depression (Wilcox 1980, Bierregaard et al. 1992, Terborgh 1992, Burkey 1995, Frankham et al. 2002, Ayres et al. 2005). Therefore, it is crucial to determine the effects of habitat fragmentation on different species to be able to concentrate conservation efforts on critical issues.

Population density is one of the most fundamental parameters influencing the dynamics of animal populations (Chiarello 2000, Efford 2004). Recent investigations on neotropical primates and small mammals in different ecosystems found in the majority of species no significant relationship between patch-size and density (Bowers and Matter 1997, Connor et al. 2000). However, in his metaanalysis, Connor et al. (2000) suggested that rare mammal species tend to have larger positive area-density correlations than more common species.

Thus a hypothesis is that rare species which are more effected by fragmentation effects than the common species should decrease in population density with decreasing fragment size while the density of common species should be unaffected or even increase in smaller forest remnants.

I investigated this hypothesis comparing the density of three rodent species in secondary forest fragments that differed in size and, furthermore, in the degree of isolation, the relation of forest edge to area and the proportion of human altered area in the vicinity of the fragment. This study forms a part of a larger project investigating the small mammal community level (Pardini et al. 2005, Umetsu and Pardini *in press*). Umetsu and Pardini (*in press*) found that the rodents *Akodon montensis* (Thomas 1902), *Oligoryzomys nigripes* (Olfers 1818) and *Delomys sublineatus* (Thomas 1903) as well as the marsupials *Gracilinanus microtarsus* (Wagner 1842) and *Marmosops incanus* (Lund 1840) differed in their response to habitat fragmentation. *D. sublineatus* was captured only in native forest and initial stages of native vegetation. Furthermore *D. sublineatus* was significantly less common in smaller and more isolated forest fragments (Pardini et al. 2005). *M. incanus* decreased in abundance in smaller and more isolated fragments and were significantly more common in native vegetation (Pardini et al. 2005, Umetsu and Pardini *in press*). *A. montensis* as well as *O. nigripes* responded in a generalist way and were captured in native forest but as well in anthropogenic altered habitats like Eucalyptus plantations and agricultural areas. *G. microtarsus* also behaved as a generalist and preferred initial stages of vegetation and was less captured in native forest vegetation (Umetsu and Pardini *in press*).

Based on these results I chose these three rodent species to further investigate the relationship between population density and habitat fragmentation of generalist and specialist rodent species. By sampling over five sampling periods and one and a half years I intend to achieve information about mean densities and the influence of different landscape parameter on the mean population densities of these species.

METHODS

INVESTIGATED SPECIES

A. montensis is a terrestrial rodent with a body mass of 19-57 g and is insectivore-omnivore. *O. nigripes* lives on the ground as well as arboreal and weighs about 9-40 g. It feeds on fruits and seeds. The biggest investigated species is the terrestrial *D. sublineatus* with 20-75 g (all information about locomotion habits, weights and feeding habits from Fonseca et al. (1996) and Emmons and Feer (1997)). Feeding habits of *D. sublineatus* are so far unknown. All three species are nocturnal and belong to the family Muridae, subfamily Sigmodontinae (Musser and Carleton 1993). *D. sublineatus* is endemic to the Atlantic forest (Fonseca et al. 1996, Bonvicino et al. 2002) whereas *A. montensis* and *O. nigripes* occur in other biomes as well (Nitikman and Mares 1987, Dalmagro and Vieira 2005).

The marsupials *Marmosops incanus* and *Gracilinanus microtarsus* belong to the family Didelphidae, subfamily Thylaminae (Tyndale-Biscoe 2005). *M. incanus* weighs about 50-140 g and feeds like the somewhat smaller *G. microtarsus* (19-31g) mainly on arthropods but also on fruits (Fonseca and Kierulff 1989, Fonseca et al. 1996, Martins and Bonato 2004). *M. incanus* is a scansorial species while *G. microtarsus* is arboreal. Both species are endemic to the Atlantic forest (Fonseca and Kierulff 1989, Fonseca et al. 1996).

STUDY AREA

The study was conducted in the region of Caucaia do Alto (23°40´ S, 47°01´ W), situated in the municipalities of Cotia and Ibiúna, São Paulo state, about 80 km south-west of the City of São Paulo, Brazil, in a transition zone between dense ombrophilous forest and semi-deciduous forest classified as "Lower Montane Atlantic rainforest" (Oliveira-Filho and Fontes 2000). The altitude varies between 800 and 1100 m (Ross and Moroz 1997). Monthly mean temperature ranges from a minimum of 11°C to a maximum of 27°C. Precipitation annually equals 1300-1400 mm and fluctuates seasonally with the driest and coldest months between April and August.

The area includes a fragmented area and a large, lower mountainous Atlantic forest area (Morro Grande Reserve). The fragmented area consists of secondary forest fragments embedded in an agricultural landscape. Secondary forest covers 31 % of the landscape, which is dominated by anthropogenic altered habitat (agricultural fields 33 %, areas with rural buildings or urban areas 15 %, vegetation in early stages of regeneration 10 %, pine and eucalyptus plantations 7 %, 4% others). The Morro Grande Reserve consists mainly of secondary forest as well. Only a minor part of the reserve provides mature forest. A more detailed description of the study area can be found in Pardini et al. (2005)

STUDY SITES

Five fragments outside the Morro Grande Reserve and one control area within the reserve were selected as study sites. All sites are of secondary growth forest and between 50 and 80 years of age. Two fragments are about 14 ha (S1, S2 in Table 1 and Fig. 1), another two fragments are approximately twice this large (30 ha; M1, M2 in Table 1 and Fig. 1). The fifth fragment is the largest, embracing 175 ha of secondary forest (B). To distinguish fragmentation effects from other uncontrolled parameters I chose a control site (C) which is likewise situated in secondary forest and part of the 10 000 ha sized Morro Grande Reserve. By this means, I ensured that the six study sites differed in fragmentation parameters but not in the forest's age.

LANDSCAPE VARIABLES

Landscape variables of the study sites were provided by courtesy of J.P. Metzger (Laboratório de Ecologia de Paisagens e Conservação – LEPaC, Departamento de Ecologia, Instituto de Biociências, University of São Paulo). To analyze the structure of the study area, air photographs of the year 2000 in a scale of 1/10000 were used. The six study sites were characterized by the following four parameters:

1. Size: The size of the fragment in ha. These values are ln-transformed to account for the large heterogeneity in the data.

2. Edge density: This parameter describes the influence of edges in the investigated fragment. It was calculated by the length of the forest edge in contact to non-forest in meter divided by the area of an 800 m radius circle around the center of the fragment in ha (≈ 201,06 ha). The radius of 800 m results from limitations by the aerial photographs and from the objective to minimize the overlap of circles in neighboring fragments.

3. Connectivity: The area of the fragment plus the area connected to the fragment by corridors of natural vegetation (including initial stages of succession). These values were ln-transformed to account for the large heterogeneity in the data.

4. Proportion of altered area: The area of agricultural use in a circle with an 800 m radius (measured as the sum of the pixels of agricultural area on the aerial photos) weighted by the distance to the central point of the fragment in meter. Thus, this index value is increased by more agricultural area and/or shorter distance to the central point of the fragment.

The calculations of the measurements were carried out using the programs FRAGSTATS and ArcView™ (Environmental Systems Research Institute, Inc., USA).

TRAPPING

A regular trapping grid of one hundred trap locations with trapping points 20 m apart from one another was established in all six study sites. Two live traps (Sherman Traps Inc., Tallahassee, USA) were set up at each trap location; a small and a large trap, sizing 23 x 9 x 8 cm and 38 x 11 x 10 cm, respectively. One trap was put on the ground, the other one at an approximate height of 1.0 to 1.5 m in lianas or trees, alternating the positions of small and large traps from one trap location to the next. Data collection was accomplished during five trapping sessions in each of the study sites from July 2003 to March 2005 (1st session: 23.07.-19.09.2003, 2nd session: 27.09.- 7.11.2003, 3rd session: 4.03.-7.04.2004, 4th session: 18.05.- 26.06.2004, 5th session: 27.01.-3.03.2005). Each trapping session consisted of 6 nights of capture. In total, this sums up to 6000 trap nights per study site. The traps were baited with banana and a mixture of peanut butter, oat and sardines. Traps were checked every morning and rebaited if necessary. Captured animals were anaesthetised for 30-60 sec (Forene®, Abbott GmbH, Wiesbaden, Germany) and marked individually by numbered ear tags (Fish and small animal tag size 1, National

Band and Tag Co., Newport, Kentucky, USA). In addition to sexing and weighing, the length of their tibia was measured to the nearest 0.5 mm. All individuals were released at their respective trapping location on the same day they were caught.

STATISTICAL ANALYSIS

The population densities were estimated by dividing the population sizes, calculated as minimum number known alive (MNKA, (Krebs 1966)), by the effective trapping area in ha. The effective trapping area was calculated by adding a boundary strip of half of the mean maximum distance moved (MMDM) to the area of the trapping grid (Otis et al. 1978, Wilson and Anderson 1985). MMDM was calculated as the mean over all study sites for the respective species. Population densities were estimated for each species and each trapping session. Subsequently the mean of the population density for

each study site was used as dependent variable in a linear regression to investigate possible relations between population densities and the landscape variables. The residuals of the regressions were tested by the Kolmogorov-Smirnov-Test and did not deviate from normality. The statistical analyses were carried out using the programs Sigma-Plot for Windows Version 8.02 and SPSS 11.5.1 (both SPSS Inc., Chicago, USA).

RESULTS

SPECIES

In total, 1074 individuals of the focus species were captured 2554 times in 36000 trap nights, leading to a mean trapping success of 7.1 % (*Akodon montensis*: 479 individuals, *Oligo-ryzomys nigripes*: 158 individuals, *Delomys sublineatus*: 153 individuals, *M. incanus*: 144 individuals, *G. microtarsus*: 140 individuals).

Tab. 3 Mean densities of *A. montensis, O. nigripes, D. sublineatus, M. incanus.* and *G. microtarsus* in the different study sites ± SE. S1, S2: small study sites; M1, M2: medium sized study sites; B: big study site; C: control site in the Morro Grande Reserve

Species	Mean Density					
	S1	S2	S3	S4	S5	CS
A. montensis	3,59 ± 1,38	1,46 ± 0,49	8,92 ± 2,18	8,63 ± 2,06	2,50 ± 0,85	4,06 ± 1,15
O. nigripes	1.00 ± 0,58	0,14 ± 0,09	2,93 ± 1,69	2,35 ± 1,22	0,41 ± 0,18	0,72 ± 0,45
D. sublineatus	1,48 ± 0,33	1,14 ± 0,55	0,37 ± 0,23	0,38 ± 0,19	1,18 ± 0,51	2,96 ± 0,61
M. incanus	1,92 ± 0,37	0,96 ± 0,12	1,03 ± 0,36	0,96 ± 0,34	0,92 ± 0,18	0,59 ± 0,15
G. microtarsus	1,23 ± 0,14	-	1,56 ± 0,70	1,31 ± 0,39	0,41 ± 0,21	0,64 ± 0,17

Additionally, I captured the didelphid marsupials *Didelphis aurita* (Wied-Neuwied 1826) and *Monodelphis americana* (Müller 1776), as well as the rodents *Oryzomys russatus* (Wagner 1848), *Oryzomys angouya* (Fischer 1814), *Thaptomys nigrita* (Lichtenstein 1829), and *Brucepattersonius* aff. *iheringi* (Thomas 1896). These species were only captured occasionally and therefore excluded from further analysis. Identification of the species was based on voucher species collected by Renata Pardini in the same region which are located in the Museo de Zoologia da Universidade de São Paulo (MZUSP).

MEAN MAXIMUM DISTANCE MOVED

Mean maximum distance moved MMDM did not differ significantly between study sites (chi-square < 7.4, df = 4, p > 0.05) or trapping sessions (all chi-square < 6.7, df = 5, p > 0.05) for any of the species (Table 4). Thus, the mean over all study sites for the respective species was used to calculate the effective trapping area.

MEAN DENSITIES

Mean densities varied strongly between species and also study sites. (Table 3). *A. montensis* showed highest mean densities in all forest fragments, followed by *O. nigripes*. Population densities of *D. sublineatus* and the marsupial species were much lower compared to the two other rodents.

POPULATION DENSITIES AND LANDSCAPE VARIABLES

The linear regression analysis between the landscape variables and the population density revealed different patterns for the three investigated species (Fig. 1). The population densities of *A. montensis* as well as of *O. nigripes* were not significantly associated to any of the four landscape variables (Size: $R^2_{A.mon}$ = 0.016, P = 0.812; $R^2_{O.nig}$ = 0.060, P = 0.641; Edge density: $R^2_{A.mon}$ = 0.009, P =

Tab. 4 Mean maximum distance moved (MMDM) ± SE for the species. The MMDM was calculated from the distances moved by individuals captured more than once within one capture-session.

Species	N	MMDM ± SE
A. montensis	392	26.28 ± 1.37
O. nigripes	63	30.71 ± 3.18
D. sublineatus	83	38.14 ± 3.12
M. incanus	88	53.89 ± 3.16
G. microtarsus	67	72.69 ± 5.98

0.856; $R^2_{O.nig}$ = 0.031, P = 0.737; Connectivity: $R^2_{A.mon}$ = 0.019, P = 0.795; $R^2_{O.nig}$ = 0.083, P = 0.580; Proportion of altered area: $R^2_{A.mon}$ = 0.170, P = 0.417; $R^2_{O.nig}$ = 0.304, P = 0.256, Fig. 8a and 1c). However, the population densities of *D. sublineatus* in the different study sites

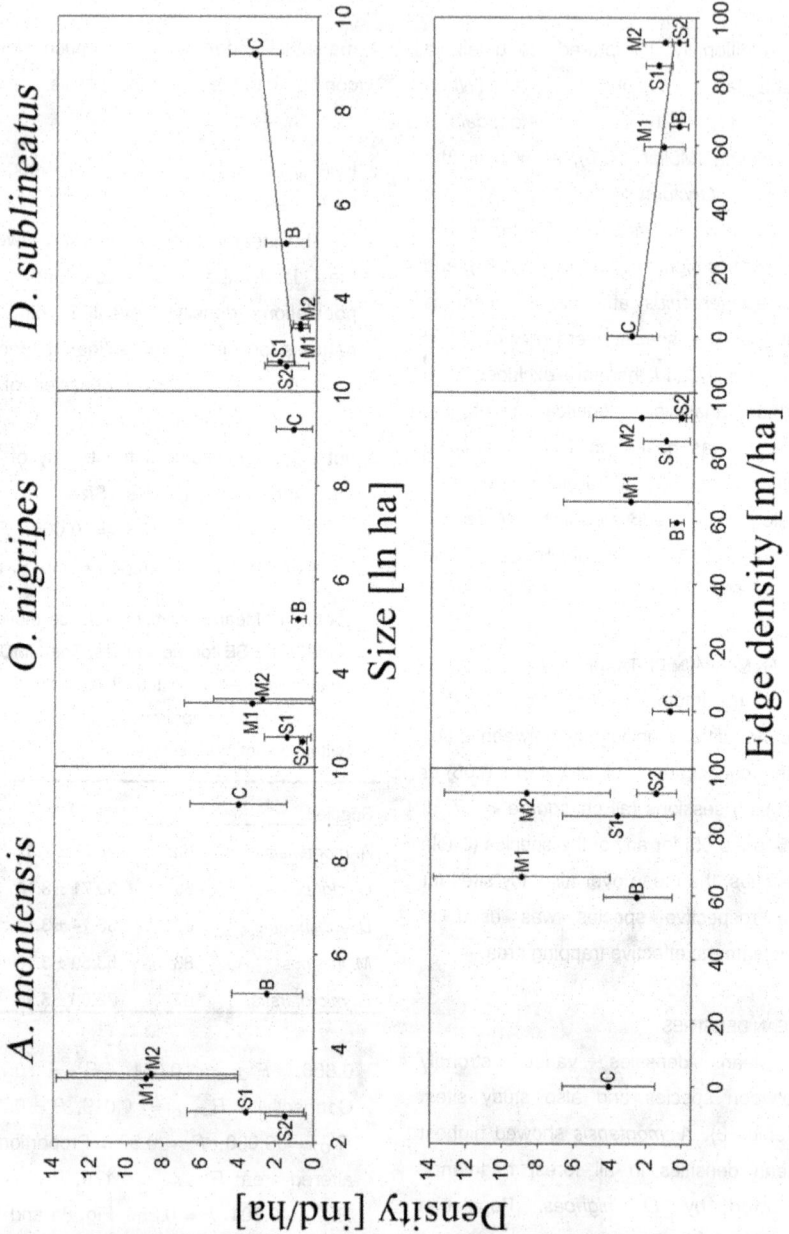

Fig. 8a. Linear regression of population densities of *A. montensis*, *O. nigripes* and D. *sublineatus* and landscape variables (*Size*, *Edge density*). Labels refer to fragment size: S1, S2: small; M1, M2: medium; B: big; C: control. See text for details.

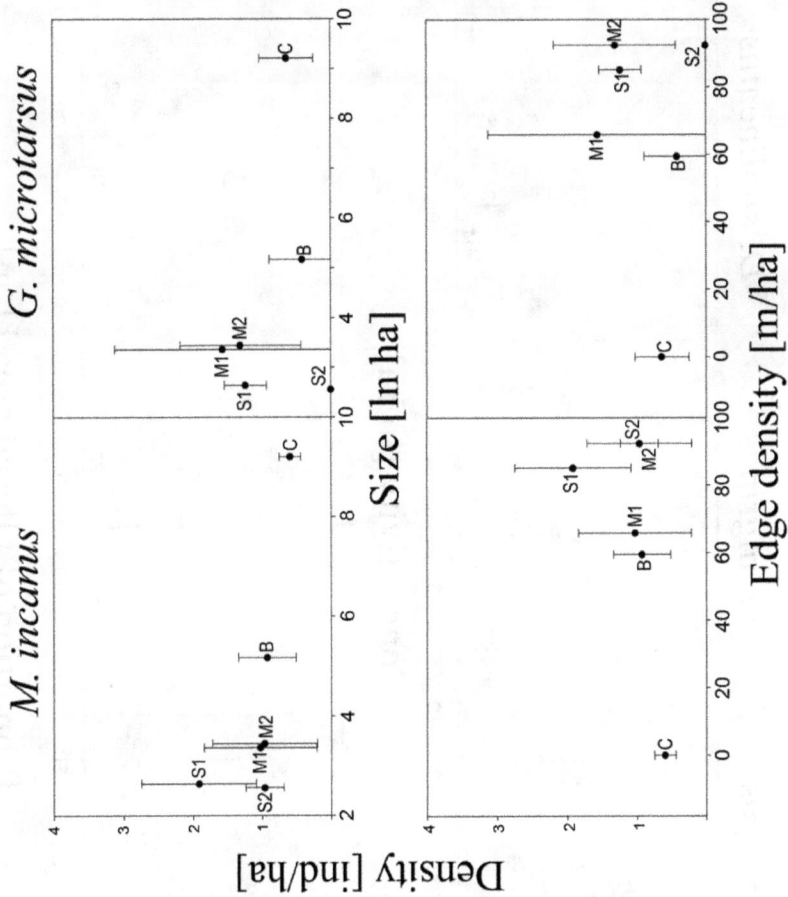

Fig. 8b Linear regression of population densities of *M. incanus and G. microtarsus* and landscape variables (*Size , Edge density*). Labels refer to fragment size: S1, S2: small; M1, M2: medium; B: big; C: control. See text for details.

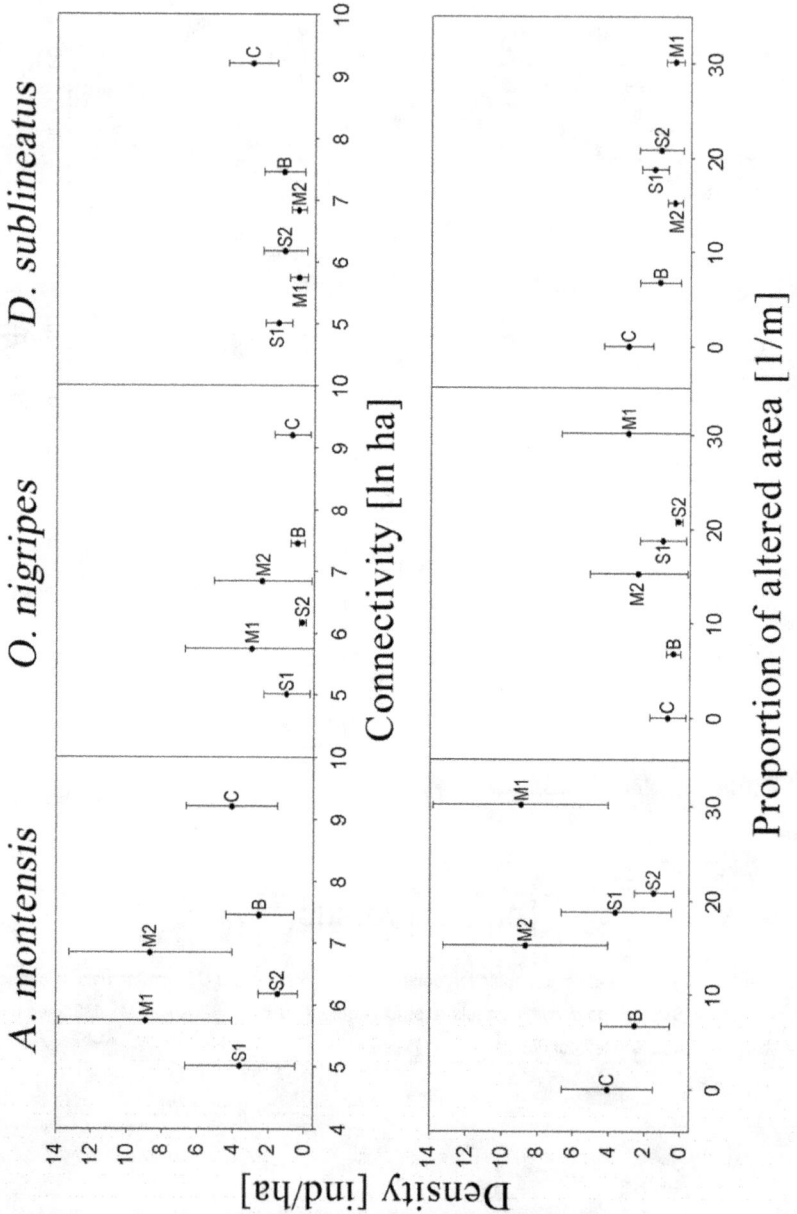

Fig. 8c Linear regression of population densities of *A. montensis*, *O. nigripes* and D. *sublineatus* and landscape variables (*Connectivity, Proportion of altered area*). Labels refer to fragment size: S1, S2: small; M1, M2: medium; B: big; C: control. See text for details.

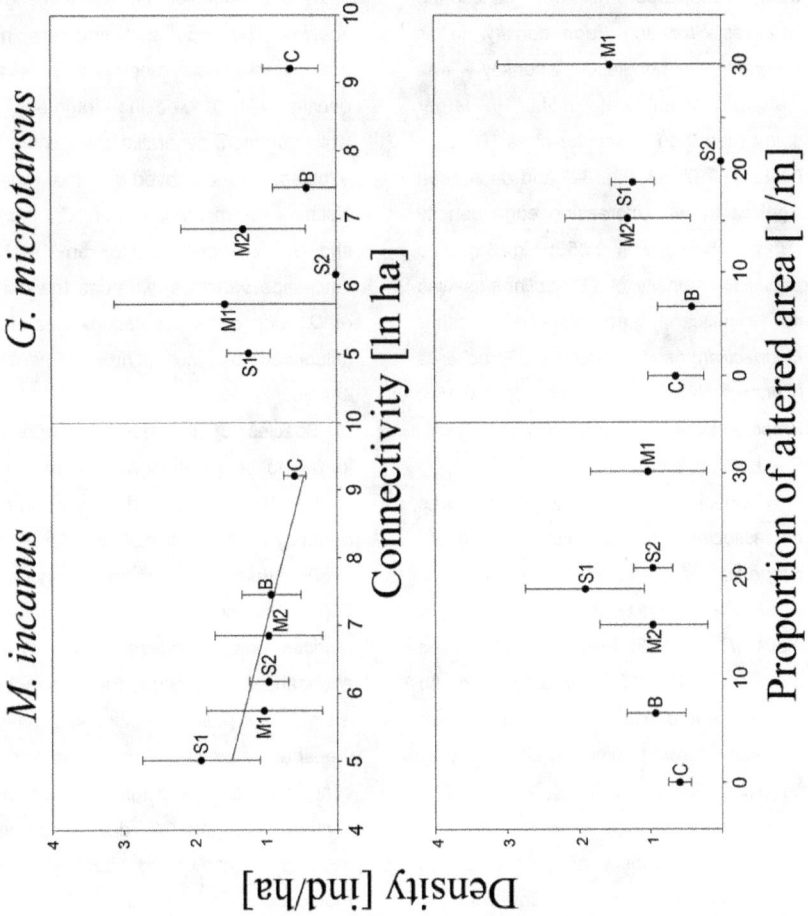

Fig. 8d Linear regression of population densities of *M. incanus and G. microtarsus* and landscape variables (*Connectivity, Proportion of altered area*). Labels refer to fragment size: S1, S2: small; M1, M2: medium; B: big; C: control. See text for details.

were associated to the landscape variables *Size* and *Edge density*. In *D. sublineatus*, population density was marginally significantly higher in larger study sites than in smaller ones ($R^2_{D.sub}$ = 0.648, P = 0.053, Fig. 8a) and decreased significantly with increasing edge density ($R^2_{D.sub}$ = 0.642, P = 0.055; Fig. 8a). The population density of *D. sublineatus* was not significantly associated to either *Connectivity* or *Proportion of altered area* (Connectivity: $R^2_{D.sub}$ = 0.432, P = 0.156; Proportion of altered area: $R^2_{D.sub}$ = 0.571, P = 0.082, Fig. 8c).

Population density of *M. incanus* was not associated to *Size*, *Edge density* or *Proportion of altered area* (Size: $R^2_{M.inc}$ = 0.395, P = 0.181; Edge density: $R^2_{M.inc}$ = 0.301, P = 0.259; Proportion of altered area: $R^2_{M.inc}$ = 0.185, P = 0.394, Fig. 8b and 1d) but density was higher in less connected forest fragments, although only marginally significant. ($R^2_{M.inc}$ = 0.650, P = 0.053, Fig. 8d).

Population density of *G. microtarsus* was not associated to any of the landscape variables (Size: $R^2_{G.mic}$ = 0.046, P = 0.682; Edge density: $R^2_{G.mic}$ = 0.017, P = 0.807; Connectivity: $R^2_{G.mic}$ = 0.142, P = 0.461; Proportion of altered area: $R^2_{G.mic}$ = 0.194, P = 0.382, Fig. 8b and 8d).

DISCUSSION

The results of this study revealed distinct differences in the influence of the landscape variables on the investigated species. The hypothesis that rare, more specialized species decline in population density with decreasing fragment size while common, generalist species are less effected was approved for the rodents. Neither *A. montensis* nor *O. nigripes* showed associations to any of the landscape variables, whereas the density of *D. sublineatus* populations was clearly influenced by increasing fragmentation effects.

Species of the genus *Akodon* are known to be generalistic with respect to habitat requirements and were captured in a variety of different habitats (Mares et al. 1989, Pardini 2004, Dalmagro and Vieira 2005). Pardini et al. (2005) captured *O. nigripes* and *A. montensis* also in high abundances in our study region and found a significant positive relationship between the abundance of *A. montensis* and the connectivity of small fragments. I could not confirm this, probably due to a longer trapping period. The density of *A. montensis* was variable during the trapping period, especially in the medium and smaller fragments as indicated by the high standard errors (Fig. 8a and 8c). Probably this is one reason why the mean density over all was not related to connectivity as shown in Pardini et al. (2005).

Oligoryzomys is also known to be able to occupy different habitats and act in a generalist way according the habitat requirements (Alho 1982, Alho et al. 1986, Nitikman and Mares 1987, Mares et al. 1989, Stallings 1989, Pardini 2004). According to the results of Umetsu and Pardini (*in press*) it is likely that *A. montensis* and *O. nigripes* are able to cross the matrix and (re)colonize small and isolated forest remnants. An important predictor for the persistence of a species in a fragmented landscape is its ability to utilize the matrix between fragments (Laurance 1991, Andrén 1994, Krohne 1997). This seems to be the case for to the investigated rodent species. *A. montensis* and *O. nigripes*, are able to maintain populations even in smaller and isolated fragments and the fragmentation of the habitat has not an obvious effect on the density of these species. It is more likely, that other factors not investigated here have a much stronger effect on the population density compared to the fragmentation variables used in this study. As habitat generalists, are not effected by increasing fragmentation and seem to be able to build a metapopulation, although data about movement rates between fragments are not known.

Little is known about *D. sublineatus*. In our study, density not only decreases with decreasing size but also with increasing edge density (Fig. 8a). Bowman et al. (2002) suggested that a positive relation between patch size and population density might be caused by negative edge effects.

If birth rates decrease and/or death rates increase at forest edges, population density will decrease with increasing edge density in smaller and isolated fragments. Furthermore, immigration from other forest fragments plays an important role in the persistence of species in isolated fragments (Bowman et al. 2002) and can put edge effects into perspective. Low immigration rates combined with negative edge effects result in lower population density in small and isolated fragments. Larger fragments might be easier to locate for dispersing individuals compared to smaller fragments (Bowers and Matter 1997). However, the great majority of Atlantic forest small mammal species do not occur in open habitat (Stevens and Husband 1998) and inter-fragment movements between forest fragments surrounded by open areas are low (Pires et al. 2002). *D. sublineatus* is a forest species (Fonseca et al. 1996, Emmons and Feer 1997) and it is unlikely that movements between fragments by trespassing agricultural fields are frequent, although this was not tested here.

The situation in the marsupial species is less clear. Only population density of *M. incanus* showed an association with the connectivity of forest fragments to other fragments. Density was higher in less connected fragments, which was unexpected given the hypothesis of more specialized species being more vulnerable to fragmentation. *M. incanus* is found in older forest (Chapter 3) and is considered to be more specialized compared to *G. microtarsus*. In a former study, *M. incanus* decreased in abundance in smaller and more isolated fragments (Pardini et al. 2005). This could not be approved in this study. Reasons for this remain unclear and it seems likely that other factors, for example food availability, might have an influence on population density of this species. Reasons for the variability of *G. microtarsus* also remain obscure. *G. microtarsus* was not captured in one of the small fragments although it was connected to other fragments.

Pardini et al. (2005) showed that the reduction of fragment size seems to influence especially the terrestrial rodents *Oryzomys russatus*, *Thaptomys nigrita* and *D. sublineatus* and that these species are the most vulnerable to Atlantic forest fragmentation. Low population densities in small fragments mean a reduced population size. A small population in turn is more vulnerable to chance extinctions and environmental-demographic stochasticity compared to large populations (Lefkovitch and Fahrig 1985). Other species like *A. montensis, O. nigripes*, or *G. microtarsus* which are less influenced by fragmentation effects, might take advantage of less interspecific competition in small fragments resulting from occasional extinction processes. Pardini et al. (2005) found a positive relationship between alpha diversity and fragment size in the study area, indicating, that the number of species is higher in larger fragments and thus potential interspecific competition might be higher as well.

Investigations on the home ranges of the species in the study area (Chapter 1) revealed the smallest movement distances for *A. montensis*, followed by *O. nigripes*, *D. sublineatus*, *M. incanus*, and *G. microtarsus*, but no significant difference in movement distance was detected between study sites for any of the species. Even the small fragments were larger than potential home range requirements (based on movement data, Chapter 1). Thus, space seems not to be a limiting factor in this case. It seems, that in our study there is no influence of the population density on movement distances and vice versa, which might be the case if food resources in small fragments would be limited.

The investigation revealed that population density of *D. sublineatus* is negatively effected by fragmentation effects while the densities of *A. montensis* and *O. nigripes* remain unaffected. The results of this study underline the importance of large forest fragments for the persistence of specialized small mammal species and ensure species' biodiversity in the Atlantic forest fragmented landscape. For the marsupial species, no clear pattern of population density regarding the landscape variables could be identified. Further studies on these species are needed to contribute substantial data for conservation concepts in this region.

REFERENCES

Alho, C. J. R. 1982. Brazilian rodents: their habitats and habits. Mammalian biology in South America 6:143-166.

Alho, C. J. R., L. A. Pereira, and A. C. Paula. 1986. Patterns of habitat utilization by small mammal populations in cerrado biome of central Brazil. Mammalia 50:447-460.

Andrén, H. 1994. Effects of habitat fragmentation on birds and mammals in landscapes with different proportions of suitable habitat: a review. Oikos 71:355-366.

Ayres, J. M., G. A. B. Fonseca, A. B. Rylands, H. L. Queiroz, L. P. Pinto, D. Masterson, and R. B. Cavalcanti. 2005. Os corredores ecológicos das florestas tropicais do Brasil. Sociedade Civil Mamariauá, Belém.

Bierregaard, R. O., T. E. Lovejoy, K. Valerie, A. A. dos Santos, and R. W. Hutchings. 1992. The biological dynamics of tropical Rainforest fragments: a prospective comparison of fragments and continuous forest. BioScience 42:859-866.

Bonvicino, C. R., S. M. Lindbergh, and L. S. Maroja. 2002. Small non-flying mammals from conserved and altered areas of Atlantic Forest and Cerrado: comments on their potential use for monitoring environment. Brazilian Journal of Biology 62:765-774.

Bowers, M. A., and S. F. Matter. 1997. Landscape Ecology of mammals: relationships between density and patch size. Journal of Mammalogy 78:999-1013.

Bowman, J., N. Cappuccino, and L. Fahrig. 2002. Patch Size and Population Density: the Effect of Immigration Behavior. Conservation Ecology 6:http://www.consecol.org/vol6/iss1/art9.

Brewer, S., and M. Reimánek. 1999. Small rodents as significant dispersers of tree seeds in a Neotropical forest. Journal of Vegetation Science 10:165-174.

Burkey, T. V. 1995. Extinction rates in archipelagoes: Implications for populations in fragmented habitats. Conservation Biology 9:527-541.

Chiarello, A. G. 2000. Density and population size of mammals in remnants of brazilian Atlantic forest. Conservation Biology 14:1649-1657.

Connor, E. F., A. C. Courtney, and J. M. Yoder. 2000. Individuals-area relationships: the relationship between animal population density and area. Ecology 81:734-748.

Dalmagro, A. D., and E. M. Vieira. 2005. Patterns of habitat utilization of small rodents in an area of Araucaria forest in Southern Brazil. Austral Ecology 30:353-362.

Efford, M. 2004. Density estimation in live-trapping studies. Oikos 106:598-610.

Emmons, L. H., and F. Feer. 1997. Neotropical rainforest mammals: a field guide. 2nd edition. The University of Chicago Press, Chicago.

Fonseca, G. A. B., G. Herrmann, Y. L. R. Leite, R. A. Mittermeier, A. B. Rylands, and J. L. Patton. 1996. Lista anotada dos mamíferos do Brasil. Occasional Papers in Conservation Biology 4:1-38.

Fonseca, G. A. B., and M. C. M. Kierulff. 1989. Biology and natural history of brazilian Atlantic forest small mammals. Bulletin of the Florida State Museum, Biological Sciences 34:99-152.

Forget, P.-M. 1991. Evidence for secondary seed dispersal by rodents in Panama. Oecologia 87:596-599.

Forget, P.-M. 1993. Post-dispersal predation and scatterhoarding of *Dipteryx panamensis* (Papilionaceae) seeds by rodents in Panama. Oecologia 94:255-261.

Forget, P.-M., T. Milleron, F. Feer, O. Henry, and G. Dubost. 2000. Effects of dispersal pattern and mammalian herbivores on seedling recruitment for *Virola mechelii* (Myristicaceae) in French Guiana. Biotropica 32:452-462.

Frankham, R., J. D. Ballou, and D. A. Briscoe. 2002. Introduction to Conservation Genetics. Cambridge University Press, Cambridge.

Galindo-Leal, C., and I. de Gusmão Câmara. 2003. The Atlantic forest of South America: Biodiversity status, threats, and outlook. Island Press, Washington, Covelo, London.

Janos, D. P., and C. T. Sahley. 1995. Rodent dispersal of vesicular-arbuscular mycorrhizal fungi in Amazonian Peru. Ecology 76:1852-1858.

Krebs, C. J. 1966. Demographic changes in fluctuating populations of *Microtus californicus*. Ecological Monographs 36:239-273.

Krohne, D. T. 1997. Dynamics of metapopulations of small mammals. Journal of Mammalogy 78:1014-1026.

Laurance, W. F. 1991. Ecological correlates of extinction proneness in australian tropical rain forest mammals. Conservation Biology 5:79-89.

Lefkovitch, L. P., and L. Fahrig. 1985. Spatial characteristics of habitat patches and population survival. Ecological Modelling 30:297-308.

Mares, M. A., J. K. Braun, and D. Gettinger. 1989. Observations on the distribution and ecology of the mammals of the cerrado grassland of central Brazil. Annals of Carnegie Museum 58:1-60.

Martins, E. G., and V. Bonato. 2004. On the diet of *Gracilinanus microtarsus* (Marsupialia, Didelphidae) in an Atlantic rainforest fragment in southeastern brazil. Mammalian Biology 69:58-60.

Musser, G. G., and M. D. Carleton. 1993. Family Muridae. Pages 501-753 *in* D. E. Wilson and D. M. Reeder, editors. Mammal species of the world: a taxonomic and geographic reference. Smithsonian Institution Press, Washington and London.

Myers, N., R. A. Mittermeier, C. G. Mittermeier, G. A. B. Fonseca, and J. Kent. 2000. Biodiversity hotspots for conservation priorities. Nature 403:853-858.

Nitikman, L. Z., and M. A. Mares. 1987. Ecology of small mammals in a gallery forest of Central Brazil. Annals of Carnegie Museum 56:75-95.

Oliveira-Filho, A. T., and M. A. L. Fontes. 2000. Patterns of floristic differentiation among Atlantic Forests in Southeastern Brazil and the influence of climate. Biotropica 32:793-810.

Otis, D. L., K. P. Burnham, G. C. White, and D. R. Anderson. 1978. Statistical inference from capture data on closed animal populations. Wildlife Monographs 62:5-135.

Pardini, R. 2004. Effects of forest fragmentation on small mammals in an Atlantic Forest landscape. Biodiversity and Conservation 13:2567-2586.

Pardini, R., S. Marques de Souza, R. Braga-Neto, and J. P. Metzger. 2005. The role of forest structure, fragment size and corridors in maintaining small mammal abundance and diversity in an Atlantic forest landscape. Biological Conservation 124:253-266.

Pimentel, D. S., and M. Tabarelli. 2004. Seed dispersal of the palm *Attalea oleifera* in a remnant of the brazilian Atlantic forest. Biotropica 36:74-84.

Pires, A. S., and F. A. S. Fernandez. 1999. Use of space by the marsupial *Micoureus demerarae* in small Atlantic forest fragments in southeastern Brazil. Journal of Tropical Ecology 15:279-290.

Pires, A. S., P. Koeler Lira, F. A. S. Fernandez, G. M. Schittini, and L. C. Oliveira. 2002. Frequency of movements of small mammals among Atlantic coastal forest fragments in Brazil. Biological Conservation 108:229-237.

Pizo, M. A. 1997. Seed dispersal and predation in two populations of *Cabralea canjerana* (Meliaceae) in the Atlantic forest of southeastern Brazil. Journal of Tropical Ecology 13:559-578.

Ross, J. L. S., and I. C. Moroz. 1997. Mapa Geomorfológico do Estado de São Paulo: escala 1:500.000. FFLCH-USP. IPT and Fapesp, São Paulo.

Sánchez-Cordero, V., and R. Martínez-Gallardo. 1998. Postdispersal fruit and seed removal by forest-dwelling rodents in a lowland rainforest in Mexico. Journal of Tropical Ecology 14:139-151.

Stallings, J. R. 1989. Small mammal inventories in an eastern brazilian park. Bulletin of the Florida State Museum, Biological Sciences 34:159-200.

Stevens, S. M., and T. P. Husband. 1998. The influence of edge on small mammals: evidence from Brazilian Atlantic forest fragments. Biological Conservation 85:1-8.

Tabarelli, M., L. P. Pinto, J. M. C. Silva, M. Hirota, and L. Bedê. 2005. Challenges and Opportunities for Biodiversity Conservation in the Brazilian Atlantic Forest. Conservation Biology 19:695-700.

Terborgh, J. 1992. Maintenance of diversity in tropical forests. Biotropica 24:283-292.

Tyndale-Biscoe, H. 2005. Life of Marsupials. CSIRO Publishing, Collingwood.

Umetsu, F., and R. Pardini. *in press*. Small mammals in a mosaic of forest remnants and anthropogenic habitats - evaluating matrix quality in an Atlantic forest landscape. Landscape Ecology.

Vieira, E. M., and P. Izar. 1999. Interactions between aroids and arboreal mammals in the brazilian Atlantic rainforest. Plant Ecology 145:75-82.

Vieira, E. M., M. A. Pizo, and P. Izar. 2003. Fruit and seed exploitation by small rodents of the Brazilian Atlantic forest. Mammalia 67:533-539.

Viveiros de Castro, E. B., and F. A. S. Fernandez. 2004. Determinants of differential extinction vulnerabilities of small mammals in Atlantic forest fragments in Brazil. Biological Conservation 119:73-80.

Wilcox, B. A. 1980. Insular Ecology and Conservation. Pages 95-117 *in* M. E. Soulé and B. A. Wilcox, editors. Conservation Biology: An Evolutionary-Ecological Perspective. Sinauer Associates, Sunderland, Massachusetts.

Wilson, K. R., and D. R. Anderson. 1985. Evaluation of two density estimators of small mammal population size. Journal of Mammalogy 66:13-21.

CHAPTER 3

Responses of five small mammal species to differences in vegetation structure in secondary Atlantic forest remnants

ABSTRACT

The Brazilian Atlantic forest is highly endangered and only about 7% of the original forest remains. The remnants consist in large parts of secondary forest and are an important habitat for a high number of endemic species. I examined the utilization of certain vegetation structures by generalist and specialist rodents and marsupials (Rodents: *Akodon montensis* (Thomas 1902), *Oligoryzomys nigripes* (Olfers 1818), *Delomys sublineatus* (Thomas 1903); Marsupials: *Marmosops incanus* (Lund 1840), *Gracilinanus microtarsus* (Wagner 1842) in six secondary forest remnants of the Atlantic forest. Furthermore, I investigated the influence of fragment sizes on the strength of preference performed by the species. *A. montensis* and *O. nigripes* used locations with low canopy and a dense understory. *M. incanus* and *D. sublineatus* were captured in locations with a closed canopy while *G. microtarsus* used locations with open canopy. The strength of preference was negatively correlated with fragment size and positively with the availability of preferred vegetation structure in the generalists *A. montensis* and *O. nigripes*. No significant correlation was found in *M. incanus*, *D. sublineatus* and *G. microtarsus*. The results indicated that generalists were able to benefit from fragmentation effects in terms of optimized utilization of vegetation structure. Furthermore the results underline the importance of mature forest structures as an adequate habitat for specialized Atlantic forest species.

INTRODUCTION

The coastal Atlantic forest is one of the most diverse but at the same time one of the most threatened natural environments in the world (Myers *et al.* 2000). Due to severe human impact over the last few centuries most of the primary coastal Atlantic forest has been destroyed. Only 7% of its original extent yet remains (Myers *et al.* 2000). This does not only include primary forest, but also secondary forest remnants in different stages of regeneration (Galindo-Leal & Câmara 2003, Dunn 2004). The stage of

regeneration of a forest and the degree of fragmentation influence the forest structure, which can determine habitat suitability for certain species and affect their occurrence as well as the composition of animal communities (Tews *et al.* 2004). Thus, secondary forests may play an important role in species conservation if primary forest habitats are limited (Dunn 2004).

Small mammals play a key role in forest ecosystems in terms of seed dispersal (Forget 1991, Sánchez-Cordero and Martinez-Gallardo 1998, Brewer and Reijmanek 1999, Vieira and Izar 1999,

Vieira *et al.* 2003, Pimentel and Tabarelli 2004) and the dispersal of mycorrhizal fungi (Janos *et al.* 1995). Furthermore, they affect the forest's regeneration by predation on seedlings (Forget 1993, Pizo 1997, Sánchez-Cordero and Martinez-Gallardo 1998, Vieira *et al.* 2003). Small mammals are impaired by fragmentation with varying intensity. Some species are able to persist in isolated fragments and cross or even occupy non-native vegetation in human-altered areas while other more specialized species are more restricted to large and/or connected forest remnants (Pires *et al.* 2002, Pardini 2004, Viveiros de Castro and Fernandez 2004, Pardini *et al.* 2005, Umetsu and Pardini, in press).

Up to now, only a limited number of studies on microhabitat use of small mammals have been carried out in secondary Atlantic forest fragments (Fonseca and Robinson 1990, Gentile and Fernandez 1999, Pardini *et al.* 2005). These studies revealed consistently that rodent and marsupial species differed in their responses to forest structure and forest fragmentation. Fonseca and Robinson (1990) as well as Gentile and Fernandez (1999) were able to distinguish microhabitat preferences between different small mammal species in secondary forest. In a recent study on small mammals in Caucaia do Alto, São Paulo, Pardini *et al.* (2005) showed that three rodent and two marsupial species were

influenced in different ways by fragmentation. The rodents *Akodon montensis* (Thomas 1902) and *Oligoryzomys nigripes* (Olfers 1818) as well as the marsupial *Gracilinanus microtarsus* (Wagner 1842) responded in a generalistic way and were not effected by fragmentation while the rodent *Delomys sublineatus* (Thomas 1903) and the marsupial *Marmosops incanus* (Lund 1840) responded as habitat specialists, decreased in abundance in smaller and more isolated fragments and were significantly more common in native vegetation (Pardini *et al.* 2005, Umetsu and Pardini, in press). Such information is needed to understand the response of the Atlantic forest species to different levels of forest regeneration and their persistence in secondary forests in human altered landscapes.

Based on the results of Pardini *et al* (2005) and Umetsu and Pardini (in press) on the small mammal community in Caucaia do Alto I selected these three rodent and two marsupial species to further investigate the reasons for the different responses to fragmentation and forest degradation associated with fragmentation. In order to gain a better understanding of the basic requirements in vegetation structure which determine the occurrence of generalist and specialist species inhabiting the Atlantic secondary forest remnants I investigated the following hypotheses. First, species impaired by fragmentation should be more restricted to vegetation structure of older or less disturbed forest while species not impaired by fragmentation should be less restricted or even occur in more disturbed forest. Further, I investigated if the strength of preference is related to the size of the forest fragments and/or to the availability of vegetation structure. Generalist species should be able to use a wider range of habitat characteristics than specialists, making it possible to inhabit more impacted secondary forests remnants.

METHODS

STUDY AREA

The study was conducted in the region of Caucaia do Alto (23°40′ S, 47°01′ W) about 80 km south-west of the City of São Paulo, in a transition zone between dense ombrophilous forest and semi-deciduous forest classified as "Lower Montane Atlantic Rain Forest" (Oliveira-Filho and Fontes 2000) in São Paulo state, Brazil. The elevation varies between 800 and 1100 m (Ross and Moroz 1997). Monthly mean temperature ranges from a minimum of 11°C to a maximum of 27°C. Precipitation annually equals 1300-1400 mm and fluctuates seasonally with the driest and coldest months between April and August.

The area includes a fragmented area and a large, lower mountainous Atlantic forest area (Morro Grande Reserve). The fragmented area consists of secondary forest fragments embedded in an agricultural landscape. Secondary forest covers 31 % of the landscape, which is dominated by anthropogenic altered habitat (agricultural fields 33 %, areas with rural buildings or urban areas 15 %, vegetation in early stages of regeneration 10 %, pine and eucalyptus plantations 7 %, 4% others). A more detailed description of the study area can be found in Pardini et al. (2005).

STUDY SITES

This study forms a part of a larger project investigating the small mammal community level (Pardini et al 2005, Umetsu and Pardini, in press). In the overall project several forest fragments differing in size and connectivity to other fragments in the fragmented area as well as different plots in the continuous forest were studied. I chose five forest fragments and one study site in the continuous forest out of this pool of study sites with the objective to encompass the largest variation in size and connectivity to investigate microhabitat preferences of small mammals. All sites are of secondary growth forest and between 50 and 80 years old. Two fragments are about 14 ha, another two fragments are approximately twice as large (30 ha). The fifth fragment is the largest, embracing 175 ha of secondary forest. The control area also consists of secondary forest and is part of the almost 10 000 ha sized Morro Grande Reserve. The study sites were at least 8 km apart from each other and individual movement between them was unlikely (and not detected). However, other forest fragments were in vicinity to the study sites and inter-fragment movements between these forests were possible.

SMALL MAMMALS

Regular trapping grids of one hundred trap locations with traps 20 m apart from one another were established in all six study sites. Usually, the grid consisted of 10 x 10 trap locations. This grid was slightly changed for the two fragments (S1 and S2) due to their shape. One small and one large trap (23x9x8 cm and 38x11x10 cm, respectively; Sherman Traps Inc., Tallahassee, USA) were set up at each trap location. One trap was put on the ground, the other one at an approximate height of 1.0 to 1.5 m, alternating the positions of small and large traps from one trap location to the next.

Data collection was accomplished during two trapping sessions in each of the study sites from March to June 2004 (1rst session: 4.03.-7.04.2004, 2nd session: 18.05.- 26.06.2004). Each session consisted of six nights of capture. In total, this sums up to 2400 trap nights per study site. The traps were baited with banana, a mixture of peanut butter, oat and sardines.

Traps were left open for the night, checked every morning and rebaited if necessary. Captured animals were anaesthetized (Forene®, Abbott GmbH, Wiesbaden, Germany) for 1-2 minutes and marked individually by numbered ear tags (Fish and small animal tag size 1, National Band and Tag Co., Newport, Kentucky, USA). In addition to sexing and weighing, the length of their tibia was measured to the nearest 0.5 mm (measurements were taken for another study). All individuals were released at their respective trapping location.

VEGETATION CHARACTERISTICS

Each of the 600 trap locations were characterized with respect to vegetation structure within a five meter radius around the trap locations according to August (1983). Vegetation characteristics were canopy height (estimated with the help of a marked 5m-pole in meters and grouped into seven categories: <6 m, 6.1–8.0 m, 8.1–10.0 m, 10.1–12.0 m, 12.1–15.0 m, 15.1–20.0 m and 20.1–25 m), canopy cover (estimations were based on a scale from 1 to 4: 1 equals less than 25 % cover while 4 counts for 76-100 % cover), vegetation density at 0 to 0.5m, 0.5 to 1.5m and 1.5 to 3m (estimations were based on a scale from 1 to 5: 1 indicates 20 % of the specific strata is covered by plants, 5 indicates 81-100 % of the specific strata is covered by plants), the amount of bamboo (estimations were based on a scale from 1 to 4: 1 represents a percentage of cover of up to 25 of bamboo while 4 stands for a 76-100 percentage) and the quantity of horizontal structures as an indicator of the connectivity of the vegetation above ground (1-4 scale: 1 means up to a 25 % connection of surrounding trees by horizontal structures, 4 means a 76-100 % connection of trees by horizontal structures).

STATISTICAL ANALYSIS

The vegetation characteristics were standardized and reduced to three principal components to minimize correlation and to identify major traits of the vegetation structure (SPSS Principal Component Analysis, default settings, Varimax rotation).

The component scores deviated little from a normal distribution (Kolmogorov-Smirnov-Test). In analysis of variance, non-normality of data can be tolerated, especially when samples are large and balanced (Underwood 1997) .The assumption of homoscedasticity was tested by Levene´s test for homogeneity of variances.

I investigated the hypothesis that specialists prefer vegetation structure of older or less disturbed forest while generalists do not show any preference. The null hypothesis was that the mean values of component scores of the principal components at trap locations where species were captured (used trap locations) did not differ significantly from the mean values of component scores of the principal components obtained of trap locations where no individual was captured (unused trap locations). I included fragment size as a covariate in SPSS´s GLM procedure in order to eliminate its potential influence. Data were pooled over all study sites (600 trap locations).

In order to examine whether or not the response to vegetation structure is correlated with the size of the study site or the availability of different vegetation structures I calculated the difference between the component scores of used and unused trap locations. For this I subtracted the mean of the component scores of unused from the score of each used trap locations. This difference was used as an indicator for the strength of preference and was subsequently correlated with the size of the study sites and the mean component scores of all trap locations within the respective study site by Spearman´s rank correlation. Thereby I assumed that the respective mean of the PC´s represents the available vegetation structure in that study site (Fonseca and Robinson 1990). Further, size of the study site was correlated to the mean component scores of all trap locations within the respective study site. Non-parametric correlation was used due to the non-normal distribution of the size of the study sites.

All tests were conducted on SPSS 11.5.1 (SPSS Inc., Chicago, USA) using a significance level of 0.05. Only study sites with more than five individual captures of one species were included in the respective analysis of the utilization of vegetation structure. Only first–captured individuals of both capture sessions were considered to assure the independence of the observations (Gentile and Fernandez 1999). Within-session recaptures were not included. It was only separated between at least one capture (used) and no capture (unused) at a specific trap location (no weighting of trap locations by numbers of individuals captured). Thereby I aimed to keep data independent according to possible recaptures of individuals due to trap-happiness.

RESULTS

SPECIES CAPTURED

In total, 827 individuals belonging to 12 species were captured 1597 times in 14400 trapnights, resulting in a trap success of 11.1 %. The five most common species were the terrestrial rodents *Akodon montensis* and *Delomys sublineatus,,* the scansorial rodent *Oligoryzomys nigripes*, the scansorial marsupial *Marmosops incanus* and the arboreal marsupial *Gracilinanus*

microtarsus, accounting for 77.4 % of individuals captured. *A. montensis* was the most abundant species with 357 individuals captured (Table 5).

Tab. 5 Number of individuals captured of the investigated species (numbers of males, females and unsexed individuals are given in parenthesis).

Species	Number of individuals captured
A. montensis	357 (172/176/9)
O. nigripes	38 (24/14)
D. sublineatus	94 (42/49/3)
M. incanus	79 (42/37)
G. microtarsus	70 (27/43)

PRINCIPAL COMPONENT ANALYSIS OF THE VEGETATION CHARACTERISTICS TO IDENTIFY MAJOR TRAITS OF VEGETATION

The vegetation characteristics were reduced to three principal components (PC1, PC2, PC3) with an Eigenvalue greater than one, which explained 69.9 % of the variance in the data (Table 6). The factor loadings are a measure of the correlation between original habitat variables and the new variables (PC´s, Adler 1985). The correlation matrix (Table 6) revealed that the first principal component, which explained 33.9 % of the variance, described the canopy height and the density of the vegetation up to three meters. The second principal component (20.5 % variance explained) reflected the amount of bamboo and the number of horizontal structures while the third (15.5 % variance explained) des-cribed the density of the canopy (Table 6).

TEST OF HYPOTHESIS ON VEGETATION UTILIZATION BY SPECIALIST SPECIES AND GENERALISTS SPECIES

To investigate whether the five small mammal species are linked to certain vegetation structures I compared unused trap locations to used trap locations. The first principal component was negatively correlated to canopy height and positively to the density of the understory (Table 6). Considering PC1, there was a significant difference between used and unused trap locations after controlling for the effect of size for the two rodent species A. montensis and O. nigripes (A. montensis: $F = 83.78$, df $= 1$, $p < 0.0001$; O. nigripes: $F = 10.07$, df $= 1$, $p = 0.002$). Both species were captured at trap locations with a significantly higher mean value for PC1 (Fig. 9). This indicates that both species were captured frequently at locations characterized by a low canopy and dense vegetation. The difference of PC1 between used and unused trap locations was not significant for the rodent D. sublineatus

Tab. 6 Eigenvalues, variance explained and correlations (factor loadings) between the three principal components and vegetation characteristics.

	PC1	PC2	PC3
Eigenvalue	2.37	1.44	1.09
Variance explained	33.88	20.51	15.50
cumulative Variance	33.88	54.39	69.89
Canopy cover	-0.110	-0.091	0.955
Canopy height	-0.687	0.021	0.238
Vegetation density 0 - 0.5m	0.800	-0.234	-0.048
Vegetation density 0.5 - 1.5m	0.874	0.127	-0.030
Vegetation density 1.5 - 3m	0.694	0.423	0.175
Amount of Bamboo	0.027	0.749	0.083
Amount of horizontal structures	-0.016	0.784	-0.276

and the marsupials *M. incanus* and *G. microtarsus* (*D. sublineatus*: F = 0.11, df = 1, p = 0.735; M. incanus: F = 1.75, df = 1, p = 0.187; *G. microtarsus*: F = 1.98, df = 1, p = 0.16; Fig. 9). The second principal component described the amount of bamboo and horizontal structures. The difference between the mean component scores of used and unused trap locations with respect to PC2 was not significant for any of the five investigated species after controlling for the effect of size (all p > 0.05; Fig. 9). Since none of thespecies showed any relation to this PC, it is excluded from further analysis.

The third principal component was correlated to the canopy cover. The mean component scores of PC3 were significantly lower in used trap locations compared to unused trap locations for *G. microtarsus* after controlling for the effect

of size (F = 4.60, df = 1, p = 0.032; Fig. 9). It was caught in locations with an open canopy. *D. sublineatus* and *M. incanus* were captured in locations with a more closed canopy and a significantly higher mean value for PC3 compared to the unused trap locations (*D. sublineatus*: F = 7.63, df = 1, p = 0.006; *M. incanus*: F = 5.30, df = 1, p = 0.022; Fig. 9). No significant difference was found for *A. montensis* (F = 0.60, df = 1, p = 0.438) and *O. nigripes* (F = 0.06, df = 1, p = 0.808).

TEST OF HYPOTHESIS ON CORRELATION BETWEEN STRENGTH OF PREFERENCE AND THE SIZE OF THE STUDY SITE AND / OR AVAILABILITY OF VEGETATION STRUCTURE

I investigated whether the use of vegetation structure was correlated with the size of the different study sites and/or

Tab. 7 Spearman correlation coefficient r_s between strength of preference and size of the study site (all species), the availability of low trees and dense understory (described by PC1) as well as the availability of a closed canopy (described by PC3). Differences in N to Table 1 are due to the fact, that in some trap locations more than one individual was captured.

Species	N	Size of the study site	Availability of PC1	Availability of PC3
A. montensis	230	-0.214*	0.346*	0.051
O. nigripes	38	-0.428*	0.397*	0.07
D. sublineatus	81	-0,077	0.019	-0.088
M. incanus	72	-0,014	0.012	0.035
G. microtarsus	58	-0,06	0.035	0.008

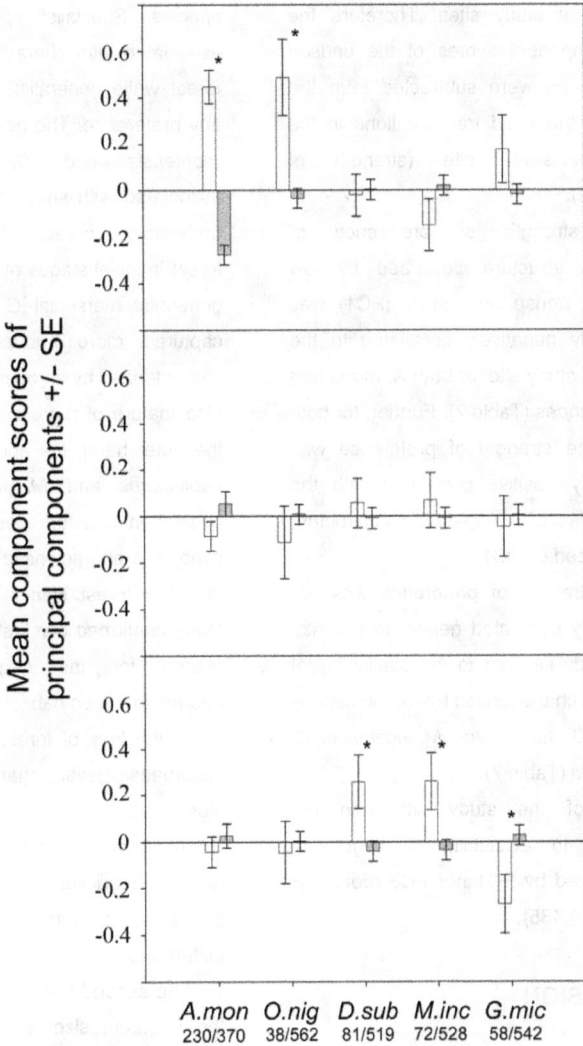

Fig. 9 Use of vegetation structure described by the three Principal Components by the different species. Comparison of mean component scores of used and unused trap locations +/- SE. Open bars represent used and grey bars unused trap locations. Asterisks mark a significant difference between used and unused trap locations. A.mon: A. montensis; O.nig: O. nigripes; D.sub: D. sublineatus; M.inc: M. incanus; G.mic: G. microtarsus. Numbers of used/unused trap locations are given below.

to the availability of vegetation structure in the different study sites. Therefore the mean component scores of the unused trap locations were subtracted from the scores of the used trap locations in the respective study site (strength of preference).

The strength of preference of vegetation structure described by low trees and dense understory (PC1) was significantly negatively correlated to the size of the study site for both *A. montensis* and *O. nigripes* (Table 7). Further, for both species the strength of preference was significantly positive correlated with the availability of vegetation structure characterized by PC1.

The strength of preference was not significantly correlated neither to the size of the study site nor to the availability of vegetation characterized by closed canopy (PC3) for D. *sublineatus*, M. *incanus* or G. *microtarsus* (Table 7).

Size of the study site was not correlated to availability of vegetation characterized by PC1 nor PC3 (both $r_s < 0.028$; $p > 0.485$).

DISCUSSION

The investigation of the response to vegetation structure of rodent and marsupial species revealed differences between the selected species. The first hypothesis was related to the question on the general preference of vegetation structure by specialist and generalist species. Specialist species should be associated with characteristics of mature forest while generalists should not show any preference. The generalist species *A. montensis* and *O. nigripes* were associated with small trees and a dense understory, characteristics of secondary forest in initial stages of regeneration. The generalist marsupial G. *microtarsus* was captured more frequently in locations characterized by an open canopy, which is also feature of more disturbed forest. On the other hand, the specialist species D. *sublineatus* and M. *incanus* preferred vegetation structure described by closed canopy, a characteristic of older and less disturbed forest. Hence, the results of our study confirmed our first hypothesis. One reason for the negative effect of fragmentation on habitat specialists seems to be the loss of forest parts with a low disturbance level, characterized in this study by a closed canopy cover. Generalist species in turn are not only able to use the disturbed habitat; they even prefer areas with a higher level of disturbance.

The second hypothesis was related to the fragment size and to availability of preferred vegetation structure. The generalist species *A. montensis* and *O. nigripes* seem to benefit from the given situation in smaller fragments since the strength of preference is negatively correlated with fragment size. The smaller the forest remnant the more they occupy the locations characterized by low canopy and dense understory. The reasons for the shift of preferred vegetation structure remain speculative. Given the result that size of the study site is not correlated to the availability of vegetation structures described by PC1, it cannot be explained by simply more available preferred habitat in the small fragments. Pardini *et al.* (2005) found in a study conducted in the same area that *A. montensis* was significantly more common in forests in earlier stages of regeneration or subjected to higher levels of disturbance, which is confirmed by our results. Further, Pardini *et al.* (2005) found that small mammal total abundance and alpha diversity decreased with decreasing fragment size. One can speculate that fewer species and fewer individuals might reduce competition in smaller fragments and *A. montensis* as well as *O. nigripes* are able to benefit by occupying areas with dense understory. In smaller fragments the influence of the matrix is intensified by stronger influence of the forest edges. *Akodon* species are found in various habitats (Mares *et al.* 1989). Recently, Dalmagro and Vieira (2005) showed that *A. montensis* responded to habitat characteristics as a habitat generalist in Araucarian forest and *O. nigripes* was captured in open habitats like savannas and agricultural fields (Alho 1982). In our study area *O. nigripes* was

also captured in the surrounding matrix (Umetsu and Pardini, in press). Therefore it is likely, that at least *O. nigripes* is capable to cross the matrix and occupy preferred vegetation structure in smaller study sites.

Both species, *A. montensis* and *O. nigripes*, seemed to occupy similar habitats. Gentile and Fernandez (1999) found, that *Akodon cursor* and *O. nigripes* showed a significant spatial overlap, which was considered to be based on ecological differences such as temporal or dietary segregation. Lacher *et al.* (1989) found no significant difference between the microhabitats of *Akodon lasiotus* and *O. nigripes* in the cerrado. These findings are supported by our results since both species responded in a similar way to vegetation structure.

Additionally, both species showed a significant positive correlation between strength of preference and availability of the preferred vegetation structure. This means that *A. montensis* and *O. nigripes* are able to select microhabitats from the range of vegetation structure available at the site. This does not seem to be the case for the specialist species *D. sublineatus* and *M. incanus* as well as for *G. microtarsus*, which do select vegetation structure (with respect to PC3), but the relationship between the strength of preference and the availability of the utilized vegetation structure is not significant. These species are for some reason not able to take advantage of the availability of the preferred vegetation structure. Furthermore, an influence of the fragment size was not detected for any of the three species since the correlation between the strength of preference and the size of the study site was not significant.

The results of our study showed that *A. montensis*, *O. nigripes* and *G. microtarsus* as habitat generalists were associated with vegetation structure characterized by disturbed forest characteristics while the specialist species *M. incanus* and *D. sublineatus* prefer vegetation structure of mature forest. Furthermore, *A. montensis* as well as *O. nigripes* are able to benefit from better availability of preferred vegetation structure in smaller sites and improve the utilization of vegetation structure in smaller forest fragments. These results confirm that not only fragmentation but also the associated structural changes of the vegetation can lead to loss of suitable habitat for endemic habitat specialist species in the Atlantic forest. Our findings underline the importance of mature forest structure as an adequate habitat for endemic species and therefore for the persistence of these species in a human influenced landscape.

REFERENCES

Adler G. H. 1985. Habitat selection and species interactions: an experimental analysis with small mammal populations. Oikos 45: 380-390.

Alho C. J. R. 1982. Brazilian rodents: their habitats and habits. Mammalian Biology in South America 6: 143-166.

August P. V. 1983. The role of habitat complexity and heterogeneity in structuring tropical mammal communities. Ecology 64: 1495-1507.

Brewer S. and Reimánek M. 1999. Small rodents as significant dispersers of tree seeds in a Neotropical forest. Journal of Vegetation Science 10: 165-174.

Dalmagro A. D. and Vieira E. M.. 2005. Patterns of habitat utilization of small rodents in an area of Araucaria forest in Southern Brazil. Austral Ecology 30: 353-362.

Dunn, R. R. 2004. Recovery of faunal communities during tropical forest regeneration. Conservation Biology 18: 302-309.

Emmons L. H. and Feer F. (eds) 1997. Neotropical rainforest mammals: a field guide, (2nd edition) edition. The University of Chicago Press, Chicago: 1-307.

Fonseca G. A. B. and Robinson J. G. 1990. Forest size and structure: competitive and predator effects on small mammal communities. Biological Conservation 53: 265-294.

Forget P. M. 1991. Evidence for secondary seed dispersal by rodents in Panama. Oecologia 87: 596-599.

Forget P. M. 1993. Post-dispersal predation and scatterhoarding of *Dipteryx panamensis* (Papilionaceae) seeds by rodents in Panama. Oecologia 94: 255-261.

Galindo-Leal C. and Câmara I. d. G. 2003. The Atlantic Forest of South America, biodiversity status, threats, and outlook. Island Press, Washington:1-488

Gentile R. and Fernandez F. A. S. 1999. Influence of habitat structure on a streamside small mammal community in a Brazilian rural area. Mammalia 63: 29-40.

Janos D. P. and Sahley C. T. 1995. Rodent dispersal of vesicular-arbuscular mycorrhizal fungi in Amazonian Peru. Ecology 76: 1852-1858.

Lacher Jr. T. E., Mares M. A. and Alho C. J. R. 1989. The structure of a small mammal community in a central Brazilian savanna. [In:. Advances in Neotropical Mammalogy. K. H. Redford and J. F. Eisenberg, eds]. Sandhill Crane Press, Gainesville, Florida: 137-162.

Mares M. A., Braun J. K. and Gettinger D. 1989. Observations on the distribution and ecology of the mammals of the cerrado grassland of central Brazil. Annals of Carnegie Museum 58: 1-60.

Myers N., Mittermeier R. A., Mittermeier C. G., Fonseca G. A. B. and Kent J. 2000. Biodiversity hotspots for conservation priorities. Nature 403: 853-858.

Nitikman L. Z. and Mares M. A. 1987. Ecology of small mammals in a gallery forest of Central Brazil. Annals of Carnegie Museum 56: 75-95.

Oliveira-Filho A. T. and Fontes M. A. L. 2000. Patterns of floristic differentiation among Atlantic Forests in Southeastern Brazil and the influence of climate. Biotropica 32: 793-810.

Pardini R. 2004. Effects of forest fragmentation on small mammals in an Atlantic Forest landscape. Biodiversity and Conservation 13: 2567-2586.

Pardini R., Marques de Souza S., Braga-Neto R. and Metzger J. P. 2005. The role of forest structure, fragment size and corridors in maintaining small mammal abundance and diversity in an Atlantic Forest landscape. Biological Conservation 124: 253-266.

Pimentel D. S. and Tabarelli M. 2004. Seed dispersal of the palm *Attalea oleifera* in a remnant of the Brazilian Atlantic Forest. Biotropica 36: 74-84.

Pires A. S., Koeler Lira P., Fernandez F. A. S., Schittini G. M. and Oliveira L. C. 2002. Frequency of movements of small mammals among Atlantic coastal forest fragments in Brazil. Biological Conservation 108: 229-237.

Pizo M. A. 1997. Seed dispersal and predation in two populations of *Cabralea canjerana* (Meliaceae) in the Atlantic Forest of Southeastern Brazil. Journal of Tropical Ecology 13: 559-578.

Ross J. L. S. and Moroz I. C. 1997. Mapa Geomorfológico do Estado de São Paulo: escala 1:500.000. FFLCH-USP. IPT and Fapesp, São Paulo.

Sánchez-Cordero V. and Martínez-Gallardo R. 1998. Postdispersal fruit and seed removal by forest-dwelling rodents in a lowland rainforest in Mexico. Journal of Tropical Ecology 14: 139-151.

Tews J., Brose U., Grimm V., Tielbörger K., Wichmann M. C., Schwager M. and Jeltsch F. 2004. Animal species diversity driven by habitat heterogeneity/diversity: the importance of keystone structures. Journal of Biogeography 31: 79-92.

Tyndale-Biscoe H. 2005. Life of Marsupials. CSIRO Publishing, Collingwood, 464 pp.

Umetsu F. and R. Pardini. in press. Small mammals in a mosaic of forest remnants and anthropogenic habitats - evaluating matrix quality in an Atlantic Forest landscape. Landscape Ecology.

Underwood, A. J. 1997. Experiments in ecology: their logical design and interpretation using analysis of variance. Cambridge University Press, Cambridge:1- 504.

Vieira E. M. and Izar P. 1999. Interactions between aroids and arboreal mammals in the Brazilian Atlantic Rainforest. Plant Ecology 145: 75-82.

Vieira E. M., Pizo M. A. and Izar P. 2003. Fruit and seed exploitation by small rodents of the Brazilian Atlantic Forest. Mammalia 67: 533-539.

Viveiros de Castro E. B. and Fernandez F. A. S. 2004. Determinants of differential extinction vulnerabilities of small mammals in Atlantic Forest fragments in Brazil. Biological Conservation 119: 73-80.

CHAPTER 4

Effects of fragmentation on parasite burden (Nematodes) of generalist and specialist small mammal species in secondary forest fragments of the coastal Atlantic forest

ABSTRACT

Parasites are considered to play an important role in the regulation of wild animal populations. I investigated parasite burden of gastrointestinal nematodes and body condition in specialist and generalist small mammal species in secondary forest fragments in the highly endangered coastal Atlantic forest. I hypothesized that body condition decreases with increasing parasite load and that parasite burden increases with increasing fragmentation in specialist species but not in generalist species as a consequence of differing responses to fragmentation effects. Investigated species were *Akodon montensis*, *Oligoryzomys nigripes*, and *Delomys sublineatus* (rodents) and the marsupials *Marmosops incanus* and *Gracilinanus microtarsus*. Prevalence was high in all species except for the arboreal *G. microtarsus* presumably because of decreased infection probability. No correlation was found between body condition and parasite load in any of the species. Contrary to our expectations, body condition of the specialists *D. sublineatus* and *M. incanus* increased in both species with increasing fragmentation. In *D. sublineatus* parasite burden increased and body condition decreased in fragments with relatively high density probably due to improved contact rates and facilitation of infection with nematodes. In all generalist species, low or no correlation between parasite burden and fragmentation was detected suggesting little effects of fragmentation on population health.

INTRODUCTION

Parasites play an important role in natural communities and various studies indicated that they are able to control host populations in size and demography analogue to the impact of predators or limitation of resources (Anderson and May 1979; Gregory 1991; Murray et al. 1997; Poulin 1999; Hugot et al. 2001; Nunn et al. 2003). For example, studies have indicated that parasitism may lead to reduced reproduction (Murray et al. 1997; Murray et al. 1998), altered host morphology (Kristan 2002), mating

behaviour (Ehman and Scott 2002; Kavaliers et al. 2005), physiology (Kristan and Hammond 2000, 2001), and increased predation risk (Murray et al. 1997; Haukisalmi and Henttonen 2000).

Thereby, the nutritional status of the host is an important factor regarding the impact of parasitism. Malnourished host individuals with degraded condition and immunological competence are less able to defend against parasite infections (May and Anderson 1979; Poulin 1996; Koski and Scott 2001). Additionally parasites may pose substantial energetic demands on the host and may further reduce host condition (Yuill 1987; Coop and Holmes

1996; Delahay et al. 1998; Gillespie and Chapman 2006). Besides the interaction with parasite infections and food supply, the conditional status of an animal might also be influenced by intraspecific and interspecific competition (Guyer 1988; Lin and Batzli 2001; Díaz and Alonso 2003) which is a function of the density of host populations. High densities facilitate stress induced by competition (May and Anderson 1979) and increase interspecific contact rates, which in turn improve transmission circumstances for directly transmitted parasites.

Few studies have been conducted on the effects of anthropogenic habitat disturbance on parasite burden within host species populations (Gillespie et al. 2005). Tests of theoretical expectations are difficult due to many confounding parameters. Fragmentation and the associated change in habitat quality (e. g. increased edge effects) may have an influence on the conditional status of species (Díaz et al. 1999) advantaging parasite infections by a depressed immune system. Isolation of host populations by increasing fragmentation of suitable habitat makes colonisation difficult for vector-driven pathogens or parasites. On the other hand, infections can have a severe impact on the isolated host population (e. g. Macdonald 1996; Allan et al. 2003).

The coastal Atlantic forest (Mata Atlântica) in Brazil is one of the most diverse and most threatened biomes in the world (Myers et al. 2000). Due to severe human impact over the last few centuries most of the primary coastal Atlantic forest has been destroyed. Only 8% of its original extent yet remains and the remaining forest is highly fragmented (Terborgh 1992; Galindo-Leal and de Gusmão Câmara 2003; Tabarelli et al. 2005). The Mata Atlântica harbours a large number of endemic species (Ayres et al. 2005). Among the 92 species of small mammals known to inhabit the Mata Atlântica, 43 are restricted to only this biome (Fonseca et al. 1996). Especially small mammals play a key role in forest ecosystems in terms of seed dispersal (Forget 1991; Sánchez-Cordero and Martínez-Gallardo 1998; Brewer and Reimánek 1999; Vieira and Izar 1999; Vieira et al. 2003; Pimentel and Tabarelli 2004) and the dispersal of mycorrhizal fungi (Janos and Sahley 1995). Furthermore, they affect the forest's regeneration by predation on seedlings (Forget 1993; Pizo 1997; Forget et al. 2000; Vieira et al. 2003).

This study forms a part of a larger project investigating the small mammal community level (Pardini et al. 2005; Umetsu and Pardini 2006. in press). Umetsu & Pardini (2006, in press) found that the rodents Akodon montensis (Thomas 1902), Oligoryzomys nigripes

(Olfers 1818) and *Delomys sublineatus* (Thomas 1903) as well as the marsupials *Marmosops incanus* (Lund 1840) and *Gracilinanus microtarsus* (Wagner 1842) differed in their response to habitat fragmentation. *D. sublineatus* and *M. incanus* were specialized according to the use of habitat compared to the other species and were captured only in native vegetation. Furthermore, the density of *D. sublineatus* decreased with increasing fragmentation effects and both *D. sublineatus* and *M. incanus* were captured mostly in locations providing mature forest vegetation structure (Püttker et al. 2006, unpublished data). On the other hand, *A. montensis, O. nigripes*, and *G. microtarsus* were habitat generalists. They were captured in native forests but as well in anthropogenic altered habitats like Eucalyptus plantations (Umetsu and Pardini 2006, in press). Density of *A. montensis* and *O. nigripes* did not correlate with fragmentation effects (Püttker et al. 2006, unpublished data) and abundance of *G. microtarsus* did not differ between large connected and small isolated forest fragments (Pardini *et al.* 2005).

Based on these results we addressed in the present study whether fragmentation has a different impact on parasite load as well as body condition of these generalist and specialist small mammal species in secondary forest fragments. We focused on gastrointestinal nematodes, one of the most important and prevalent group of parasites (Hugot et al. 2001) which are a major cause of disease and death in humans, domestic animals and wildlife (Stear et al. 1997). Since laboratory studies have shown that malnutrition of rodents led to increased parasite survival (Bolin 1977; Gbakima 1993; Koski and Scott 2001 and references therein), we expected to find a similar pattern between nutrition, estimated via body condition, and parasite burden in the field and hypothesised that body condition of species decreases with increasing parasite load. Thereby, the consequences of fragmentation might differ between generalist and specialist species. Specialist species that suffer from fragmentation effects might be faced with suboptimal circumstances, including resource availability (Diaz et al. 1999) as well as increased environmental stress in smaller fragments (Marchand 2003). The investigated coastal Atlantic forest fragments differed in size, degree of isolation, relation of forest edge to area, and proportion of human altered area in the vicinity to the fragment. Therefore, we hypothesised that body condition decreases and parasite load concurrently increases with increasing fragmentation effects in the specialist species *D. sublineatus* and *M. incanus*. On the other hand, we assumed that condition and parasite load of generalists *A. montensis, O. nigripes,* and *G. microtarsus* are not or only marginally influenced by fragmentation. Furthermore, because higher host densities improve transmission circumstances for parasites (Haukisalmi 1996; Bush 2001) we examined the relation between parasite load and density of the different species as well as correlation between condition and density.

METHODS

INVESTIGATED SPECIES

Akodon montensis is a terrestrial, insectivore-omnivore rodent with a body mass of 19-57 g. *Oligoryzomys nigripes* lives on the ground as well as arboreal and weighs about 9-40 g. It feeds on fruits and seeds. The biggest investigated rodent is the terrestrial *D. sublineatus* with 20-75 g (all information about locomotion habits, weights and feeding habits from Fonseca et al. 1996) and (Emmons and Feer 1997). Feeding habits of *D. sublineatus* are so far unknown. All three species are nocturnal and belong to the family Muridae, subfamily Sigmodontinae (Musser and Carleton 1993). *D. sublineatus* is endemic to the Atlantic forest (Fonseca et al. 1996, Bonvicino et al. 2002) whereas *A. montensis* and *O. nigripes* occur in other biomes as well (Nitikman and Mares 1987, Dalmagro and

Vieira 2005). The marsupials *Marmosops incanus* as well as *Gracilinanus microtarsus* belong to the family Didelphidae, subfamily Thylaminae (Tyndale-Biscoe 2005). *M. incanus* weighs about 50-140 g and feeds like the somewhat smaller *G. microtarsus* (19-31g) mainly on arthropods and additionally on fruits (Fonseca and Kierulff 1989, Fonseca et al. 1996, Martins and Bonato 2004). *M. incanus* is a scansorial species while *G. microtarsus* is mostly arboreal. Both species are endemic to the Atlantic forest (Fonseca and Kierulff 1989); (Fonseca et al. 1996).

STUDY AREA

The study was conducted in the region of Caucaia do Alto (23°40´ S, 47°01´ W), situated in the municipalities of Cotia and Ibiúna, São Paulo state, about 80 km south-west of the City of São Paulo, Brazil, in a transition zone between dense ombrophilous forest and semi-deciduous forest classified as "Lower Montane Atlantic rainforest" (Oliveira-Filho and Fontes 2000). The altitude varies between 800 and 1100 m (Ross and Moroz 1997). Monthly mean temperature ranges from a minimum of 11°C to a maximum of 27°C. Precipitation annually equals 1300-1400 mm and fluctuates seasonally with the driest and coldest months between April and August.

The area includes a fragmented area and a large, lower mountainous Atlantic forest area (Morro Grande Reserve). The fragmented area consists of secondary forest fragments embedded in an agricultural landscape. Secondary forest covers 31 % of the landscape, which is dominated by anthropogenic altered habitat (agricultural fields 33 %, areas with rural buildings or urban areas 15 %, vegetation in early stages of regeneration 10 %, pine and eucalyptus plantations 7 %, 4% others). The Morro Grande Reserve consists mainly of secondary forest as well. Only a minor part of the reserve provides mature forest. A more detailed description of the study area can be found in Pardini et al. (2005).

STUDY SITES

Five fragments outside the Morro Grande Reserve and one control area within the reserve were selected as study sites. All sites are of secondary growth forest and between 50 and 80 years of age (Godoy Teixeira 2005). Two fragments are about 14 ha (S1, S2), another two fragments are approximately double-sized (30 ha; M1, M2). The fifth fragment is the largest, embracing 175 ha of secondary forest (B). To distinguish fragmentation effects from other uncontrolled parameters I chose a control site (C) which is likewise situated in secondary forest and part of the 10 000 ha

sized Morro Grande Reserve. By this I ensured that the six study sites differed in fragmentation parameters but not in the forest's age.

TRAPPING

A regular trapping grid of one hundred trap locations with trapping points 20 m apart from one another was established in all six study sites. Two live traps (Sherman Traps Inc., Tallahassee, USA) were set up at each trap location; a small and a large trap, sizing 23 x 9 x 8 cm and 38 x 11 x 10 cm, respectively. One trap was put on the ground, the other one at an approximate height of 1.0 to 1.5 m in lianas or trees, alternating the positions of small and large traps from one trap location to the next. Data collection was carried out between 23.07. and 19.09.2003. At each study site 6 nights of capture were accomplished. These sums up to 1200 trap nights per study site. The traps were baited with banana and a mixture of peanut butter, oat and sardines. Traps were checked every morning and baited if necessary. Captured animals were anaesthetized for 30-60 sec (Forene®, Abbott GmbH, Wiesbaden, Germany) and marked individually by numbered ear tags (Fish and small animal tag size 1, National Band and Tag Co., Newport, Kentucky, USA). In addition to sexing and weighing, the length of their tibia was measured to the nearest 0.5 mm. All individuals were released at their respective trapping location on the same day they were caught.

CONDITION INDEX

I used the residuals from a regression of body mass on body size (length of tibia) as an index of body condition (Díaz and Alonso 2003, Schulte-Hostedde et al. 2005). Thereafter, this is referred to as condition index (CI). Because of the different body proportions and resultant differences in condition, young individuals and pregnant females were excluded from the analysis. Only first individual captures were considered in the analysis to avoid variations in body weight due to repeated capture of an individual.

DENSITY

Population densities were estimated by dividing population size, calculated as minimum number known alive (MNKA, (Krebs 1966), by effective trapping area in ha. The effective trapping area was calculated by adding a boundary strip of half of the mean maximum distance moved (MMDM) to the area of the trapping grid (Otis et al. 1978, Wilson and Anderson 1985). MMDM was calculated as the mean over all study sites for the respective species.

PARASITOLOGICAL EXAMINATIONS

I focused our analyses of para-site burden on gastrointestinal nema-todes

due to their high prevalence (about 60% of free ranging animals are usually infected (Behnke et al. 1999)) and their impact on fitness attributes in a wide range of livestock and wild animals species (Albon et al. 2002, Stien et al. 2002).The intensity of nematode infection was investigated by counting the eggs of gastrointestinal nematodes in the animal's feces using a modification of the noninvasive McMaster flotation technique (Gordon and Whitlock 1939). Nematode eggs were assigned to morphotypes based on size and morphological characters. I used fecal egg counts (FEC, number of eggs per gram feces), which reflect the overall burden of worms and their fecundity (Stear et al. 1997), and number of different nematode infections per individual (NNI) as measurements for nematode parasitism. These measures have been widely used (e.g. Schwaiger et al. 1995, Paterson et al. 1998, Coltman et al. 1999, Harf and Sommer 2005, Meyer-Lucht and Sommer 2005). I transformed the data by taking the logarithm of FEC plus one (Stear et al. 1997). Prevalence was calculated as the number of hosts infected with one or more different parasite morphotypes divided by the number of hosts examined.

LANDSCAPE VARIABLES

Landscape variables of the study sites were provided by courtesy of J.P. Metzger (Laboratório de Ecologia de Paisagens e

Conservação – LEPaC, Departamento de Ecologia, Instituto de Biociências, University of São Paulo). To analyze the structure of the study area, air photographs of the year 2000 in a scale of 1/10000 were used. The six study sites were characterized by the following four parameters:

1. Size: The size of the fragment in ha. These values are ln-transformed to account for the large heterogeneity in the data.

2. Edge density: This parameter describes the influence of forest edges in the investigated fragment. It was calculated by the length of the forest edge in contact to non-forest in meter divided by the area of an 800 m radius circle around the center of the fragment in ha (= 201,06 ha). The radius of 800 m was chosen to minimize the overlap of circles in neighboring fragments and from limitations by the aerial photographs.

3. Connectivity: The area of the fragment plus the area connected to the fragment by corridors of natural vegetation (including initial stages of succession). These values were ln-transformed to account for the large heterogeneity in the data.

4. Proportion of altered area: The area of agricultural use in a circle with an 800 m radius (measured as the sum of the pixels of agricultural area on the aerial photos) weighted by the distance to the central

point of the fragment in meter. Thus, this index value is increased by more agricultural area and/or shorter distance to the central point of the fragment. The calculations of the measurements were carried out using the programs FRAGSTATS and ArcView[TM] (Environmental Systems Research Institute, Inc., USA).

STATISTICAL ANALYSIS

I used one-way ANOVA to compare mean CI as well as mean FEC (dependent variables) between sexes and study sites (independent variables) except in *D. sublineatus*. Non-parametric Kruskal-Wallis-Test was used to compare mean FEC between study sites because FEC proved to be heteroscedastic for the different study sites in that species. In case of *A. montensis*, data on FEC was not available for all individuals.

Hochberg´s GT2 for different group sizes was run a posteriori when significant variations were found. In case of *D. sublineatus*, I use Mann-Whitney-U-Test with Bonferroni-corrected α for pair wise comparisons of mean FEC between study sites. I used linear regression of CI on landscape parameters to examine the influence of these parameters on the condition of the species. Relation between CI and FEC was investigated by Pearson´s product moment correlation. Due to non-normal distributed data,

Spearman Rank correlation was used to investigate relationships between CI and host density, CI and NNI (number of nematode infections), FEC and landscape parameters as well as between FEC and host density. The statistical analyses were carried out using the program SPSS 11.5.1 (SPSS Inc., Chicago, USA). Because of varying sample sizes between study sites, which can result in misleading prevalence measures (Gregory and Blackburn 1991, Decker et al. 2001) I did not compare prevalence between study sites.

RESULTS

SPECIES CAPTURED

In total, 331 individuals were captured 606 times in 1200 trap nights, leading to a mean trapping success of 5.5 %. 290 individuals belonged to the focus species *A. montensis* (116 individuals), *O. nigripes* (87), *D. sublineatus* (32), *M. incanus* (35), and *G. microtarsus* (20). Other species captured were the didelphid marsupials *Didelphis aurita* (Wied-Neuwied, 1826) and *Monodelphis americana* (Müller, 1776), as well as the rodents *Oryzomys russatus* (Wagner 1848), *Oryzomys angouya* (Fischer, 1814), *Thaptomys nigrita* (Lichtenstein, 1829), and *Brucepattersonius* aff. *iheringi* (Thomas, 1896). These species were only captured occasionally and therefore excluded from further analysis.

Identification of the species was based on voucher species of the Museo de Zoologia da Universidade de São Paulo (MZUSP) collected by Renata Pardini in the same region.

PARASITE LOAD

Eight different nematode morphotypes could be distinguished in the investigated host individuals. None of the investigated species was host to all nematode morphotypes (Fig. 10). However, seven of the eight morphotypes were found in *A. montensis*. One nematode morphotype was detected in all investigated species and was the most abundant helminth in all species. Another nematode was found in all species but *O. nigripes* (Fig. 10). Furthermore, two cestode morphotypes (one morphotype occurred in 27 *A. montensis* and in two *O. nigripes*; another was only found in one *D. sublineatus*) and one trematode (in eight individuals of *A. montensis* and one individual of *D. sublineatus*) were detected.

Prevalence was high in all species (*A. montensis*: 94 %, n = 73; *O. nigripes*: 100 %, n = 75; *D. sublineatus*: 97 %, n = 30, *M. incanus*: 97 %, n = 31) except for *G. microtarsus* (44.4 %, n = 18). Infected rodents carried between one (66.8 %), two (29.7 %) and three (3.4 %) nematode morphotypes, marsupials only between one (85.0 %) and two (15.0 %).

CONDITION INDEX

A. montensis

The CI differed between sexes (df = 1; F = 4.850; p = 0.031). Males had a larger mean CI compared to females. For males no difference between study sites was detected for CI (df = 5; F = 1.329; p = 0.275). None of the regression analysis between landscape parameters and CI revealed a significant correlation but males showed a tendency to have a larger CI in less connected study sites (Table 8). CI was not related to density (n = 40; r_s = -0.146; p = 0.370), FEC (n = 40; r = -0.006; p = 0.972) or NNI (n = 40; r_s = -0.033; p = 0.842). In females, CI did not differ between study sites (df = 4; F = 1.518; p = 0.225) and was not correlated to any of the landscape parameters (Table 8). No correlation was detected between CI and density (n = 32; r_s = -0.145; p = 0.429), FEC (n = 32; r = -0.159; p = 0.384) or NNI (n = 32; r_s = -0.154; p = 0.399).

O. nigripes

CI did not differ neither between sexes (df = 1; F = 1.009; p = 0.3195) nor study sites (df = 5; F = 1.149; p = 0.344) in O. nigripes. Regression analysis between landscape parameters and CI revealed no correlation (Table 8). CI did not correlate with density (n = 69; r_s = 0.032; p = 0.795), FEC (n = 69; r = -0.138, p = 0.259) nor NNI (n = 69; r_s = 0.165; p = 0.176).

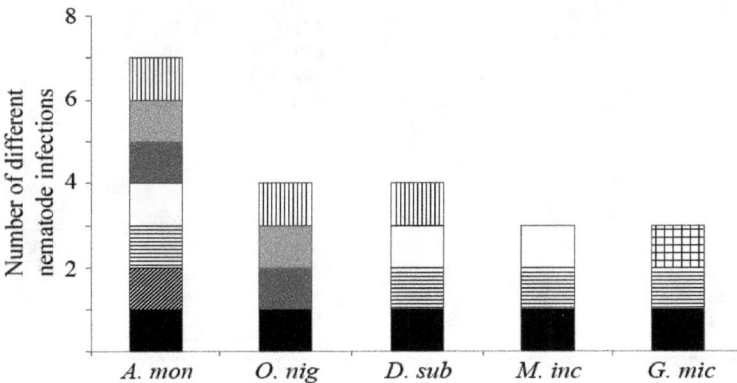

Fig. 10 Number of different nematode morphotype infections in the investigated species. Identical colour patterns refer to the same nematode morphotype. A. mon: Akodon montensis; O. nig: Oligoryzomys nigripes; D. sub: Delomys sublineatus; M. inc; Marmosops incanus; G. mic: Gracilinanus microtarsus.

Tab. 8 Slopes, intercepts, and R^2 obtained from regression analysis between the condition index (CI) and landscape parameters (size, edge density, connectivity, proportion of altered area) for the five species of small mammals. In case of differences between sexes separate values for males (m) and females (f) are given.

		Slope	Intercept	R^2	p
Size					
A. montensis	m	-0.119	0.760	0.074	0.088
	f	-0.059	-0.135	0.010	0.579
O. nigripes		-0.047	0.168	0.003	0.667
D. sublineatus		-0.133	0.812	0.179	0.028
M. incanus		-0.210	0.788	0.178	0.020
G. microtarsus		0.004	-0.013	0.000	0.995
Edge Density					
A. montensis	m	0.007	-0.211	0.049	0.168
	f	0.006	-0.787	0.023	0.406
O. nigripes		-0.002	0.117	0.001	0.814
D. sublineatus		0.010	-0.397	0.168	0.034
M. incanus		0.013	-0.966	0.128	0.052
G. microtarsus		0.045	-3.597	0.254	0.039
Connectivity					
A. montensis	m	-0.215	1.658	0.089	0.061
	f	0.008	-0.390	0.000	0.961
O. nigripes		-0.092	0.571	0.006	0.511
D. sublineatus		-0.259	1.972	0.187	0.024
M. incanus		-0.351	2.192	0.225	0.008
G. microtarsus		0.401	-2.244	0.080	0.270
Proportion of altered area					
A. montensis	m	0.020	-0.079	0.039	0.224
	f	0.000	-0.338	0.000	0.992
O. nigripes		0.009	-0.201	0.006	0.511
D. sublineatus		0.030	-0.343	0.148	0.048
M. incanus		0.035	-0.581	0.075	0.143
G. microtarsus		-0.077	1.698	0.242	0.045

D. sublineatus

The CI did not differ neither between sexes (df = 1; F = 0.698; p = 0.411) nor study sites (df = 4; F = 2.519; p = 0.07; no captures in M2, therefore limited degrees of freedom). CI correlated negatively with the landscape parameters size and connectivity and positively with edge density and proportion of altered area (Table 8). CI was not correlated with FEC (n = 27; r = -0.201, p = 0.314) or NNI (n = 27; r_s = -0.311; p = 0.114) but negatively with density (n = 27; r_s = -0.579; p = 0.002).

M. incanus

No difference in CI between sexes was detected (df = 1; F = 0.005; p = 0.944). Study sites did not differ in CI (df = 5; F = 1.705; p = 0.172). CI was negatively related to size and connectivity as well as marginally positive to edge density (Table 8). CI was correlated with density (n = 30; r_s = 0.409; p = 0.025) and not with FEC (n = 30; r = -0.113; p = 0.551) or NNI (n = 30; r_s = 0.084; p = 0.661).

G. microtarsus

CI did not differ between sexes (df = 1; F = 0.460; p = 0.508) or study sites (df = 2; F = 3.168; p = 0.073; no capture in study sites S2, B and C, therefore limited degrees of freedom). Regression analysis

revealed a positive relation between CI and edge density and a negative relation between CI and proportion of altered area (Table 8). CI was not related to size or connectivity. No correlation was found between CI and density (n = 17; r_s = 0.184; p = 0.455), FEC (n = 18; r = -0.083; p = 0.752) or NNI (n = 18; r_s = -0.349; p = 0.170).

INTENSITY OF NEMATODE INFECTION

A. montensis

FEC (Nematode) did not differ between sexes (df = 1; F = 1.940; p = 0.168) but between study sites (df = 5; F = 2.387; p = 0.047). Post hoc test tests revealed that FEC was higher in the study site B compared to study sites M1 (p = 0.048).None of the landscape parameters was correlated with FEC and FEC did not correlate with density (Table 8).

Tab. 8 Correlation matrix between parasite load (FEC-values) and landscape parameters as well as density for the five small mammal species

	A. montensis	O. nigripes	D. sublineatus	M. incanus	G. microtarsus
	FEC (n = 73)	FEC (n = 75)	FEC (n = 30)	FEC (n = 30)	FEC (n = 18)
Size	-0.044	-0.029	0.528*	0.147	0.145
Edge density	-0.016	0.341**	-0.528**	-0.577**	0.141
Connectivity	-0.033	-0.029	0.519**	-0.001	0.145
Proportion of altered area	-0.118	-0.069	-0.525**	-0.105	-0.141
Density	-0.193	-0.061	0.386**	-0.054	-0.031

O. nigripes

FEC did not differ between sexes (df = 1; F = 2.841; p = 0.096) but between study sites (df = 4; F = 3.110; p = 0.020). Post hoc test revealed that FEC was lower in study site C compared to study site M2 and S1. FEC was positively related to edge density (Table 8). No relation was found between FEC and size, connectivity as well as proportion of altered area. FEC did not correlate with density (Table 8).

D. sublineatus

FEC did not differ between sexes (df = 1; F = 0.166; p = 0.687). FEC differed between study sites (df = 4; χ^2 = 9.483; p = 0.05) but pair wise comparison revealed no difference (all Z > 0.017; all p > 0.015; Bonferroni corrected α = 0.0045). FEC was correlated positive to size and connectivity, while it was negatively correlated to edge density and proportion of altered area (Table 8). Furthermore, FEC correlated positively with density (Table 8).

M. incanus

FEC did not differ between sexes (df = 1; F = 0.936; p = 0.341) and no difference was detected between study sites (df = 5; F = 2.196; p = 0.087).

Spearman rank correlation; *p < 0.05. **p < 0.01

FEC was negatively related to edge density but not to other landscape parameters. FEC did not correlate with density (Table 8).

G. microtarsus

No difference between sexes or study sites in mean FEC was detected (sex: df = 1; F = 0.592; p = 0.453; study sites: df = 2; F = 0.608; p = 0.521). None of the landscape parameters was related to FEC and FEC did not correlate with density (Table 8).

DISCUSSION

The high prevalence of nematodes in the investigated species suggests that parasites play an important role in the dynamics of small mammal populations in the Atlantic forest. Some nematode species seem not to be very specialized because of the occurrence of the same morphotypes in different rodents and marsupials. The relatively low prevalence in G. microtarsus is presumably due to the habits of this species. In contrast to the rodents and also to the marsupial M. incanus, G. microtarsus is an aboreal species which diminishes chances to encounter feces from other individuals infested with parasite eggs or larvae.

None of the species showed a significant difference in FEC between sexes. In several studies males tend to have significantly higher parasite prevalence and intensity compared to females (reviewed in Poulin 1996; Rossin and Malizia 2002; Díaz and Alonso 2003). Morphological, physiological or behavioural reasons might be responsible for this unbalanced relation which might not be revealed during a short time interval. Long-term studies on the dynamics of parasite prevalence and the intensity of infection are necessary to investigate sexual differences in host susceptibility.

Our results did not approve our first hypothesis that condition decreases with increasing parasite load. None of the investigated species showed any relation neither between CI and FEC nor between CI and NNI. There are two possible explanations for this result. Firstly, the influence of nematode infection on body condition might be compensated by increased food intake (Tripet and Richner 1997). We did not investigate food intake rates but the investigated species are probably able to face potential energy demands caused by parasite infection by increased ingestion. Under natural conditions it is difficult to test this hypothesis since it is necessary to keep track of the amount of food ingested by the individuals, which is impossible to achieve in the field. However, it would be still informative to investigate food intake rates in correlation to parasite burden under laboratory conditions. Individuals with high parasite load should consume more food compared to individuals with low parasite load. Secondly, nematode infection might have little or no influence on the condition measured by an index of body mass/ body length in the investigated species. Kristan and Hammond (2000) found less body fat, increased metabolism, decreased glucose uptake ability in the intestine and changes in organ masses in nematode parasitized mice but observed no change in body mass. There may be for instance metabolic changes due to nematode infection which could not be detected by measuring a condition index like body mass relative to body size. Investigations on metabolism, food acquisition, and organ masses would be necessary to examine such probable consequences of nematode infection on the conditional status of the host.

In relation to the second hypothesis, the results showed clearly that the condition index of the specialist species *D. sublineatus* and *M. incanus* was more influenced by fragmentation than in the remaining investigated species. However, we expected a decrease in condition with increasing fragmentation. Instead of that both species showed an increase of the condition index in smaller and more isolated fragments and for *D. sublineatus* the FEC decreased with increasing fragmentation. In case of *D. sublineatus*, the species density seems to play an important role regarding the CI as well as the FEC. In fragments with relatively high density of *D. sublineatus*, the individuals showed low CI as well as in larger, connected fragments. Inversely FEC decreased with increasing fragmentation and correlated positively with density. In a long term study, mean density of *D. sublineatus* correlated negatively with fragment size (Püttker et al. 2006, unpublished data). High host densities can enhance transmission rates of parasites by augmented contact rates between individuals and it has been shown that host density correlates positively with parasite prevalence and diversity (May and Anderson 1979; Morand and Poulin 1998; Gillespie et al. 2005; Kavaliers et al. 2005). This might be the case in *D. sublineatus*. In smaller fragments density is lower compared to larger fragments and therefore diminished contact rates make it difficult for directly transmitted nematodes to survive in the population. Additionally the higher CI of *D. sublineatus* in smaller, more isolated fragments might lead to a better immune response which exacerbates infection. However, the lack of correlation between FEC and CI suggests that there must be additional reasons than low parasite load explaining the better CI in smaller fragments. To test the importance of population density with respect to CI and FEC a separate experiment is needed. Ideally, this would include a number of replicates of small

fragments with low density and large fragments with higher densities of *D. sublineatus*. If population density really causes the observed effects in *D. sublineatus*, FEC must be constantly higher and CI constantly lower in larger fragments. The fact that in *M. incanus* a similar CI pattern was observed but a positive correlation of CI and density and no correlation between FEC and density remains illusive.

In accordance with our second hypothesis, CI and FEC were less influenced by fragmentation in the generalist species *A. montensis*, *O. nigripes* and *G. microtarsus*. In *A. montensis* and *O. nigripes*, fragmentation had also no effect on population density (Püttker et al. 2006, unpublished data). Thus, changes of the habitat seem not to have a severe influence on body condition or FEC in these generalist species underlines their ability to survive in fragmented landscapes. Likewise, Díaz et al. (1999) found no effect of fragment size on body condition of a generalist rodent (*Apodemus sylvaticus*) in a study on fragmentation effects in Spain. However, fragmentation does not seem to improve habitat conditions for these species because CI did not increase and FEC did not decrease in smaller fragments (Díaz et al. 1999).

The results of this study might partially be influenced by confounding parameters due to uncontrolled conditions in the field. Nevertheless, the study showed the different effects of fragmentation on parasite burden of generalist and more specialized species. Considering the important role parasites have on the population dynamics of their hosts, further investigations are needed to understand these complex relationships as well as the effects of anthropogenic environmental change on the parasite load of small mammals in the Atlantic forest.

REFERENCES

Albon, S. D., A. Stien, R. J. Irvine, R. Langvatn, E. Ropstad, and O. Halvorsen. 2002. The role of parasites in the dynamics of a reindeer population. Proceedings of the ROyal Society of London Series B - Biological Sciences 269:1625-1632.

Allan, B. F., F. Keesing, and R. S. Ostfeld. 2003. Effect of forest fragmentation on Lyme disease risk. Conservation Biology 17:267-272.

Anderson, R. M., and R. M. May. 1979. Population biology of infectious deseases: Part I. Nature 280:361-367.

Ayres, J. M., G. A. B. Fonseca, A. B. Rylands, H. L. Queiroz, L. P. Pinto, D. Masterson, and R. B. Cavalcanti. 2005. Os corredores ecológicos das florestas tropicais do Brasil. Sociedade Civil Mamariauá, Belém.

Bonvicino, C. R., S. M. Lindbergh, and L. S. Maroja. 2002. Small non-flying mammals from conserved and altered areas of Atlantic Forest and Cerrado: comments on their potential use for monitoring environment. Brazilian Journal of Biology 62:765-774.

Brewer, S., and M. Reimánek. 1999. Small rodents as significant dispersers of tree seeds in a Neotropical forest. Journal of Vegetation Science 10:165-174.

Coltman, D., J. Pilkington, J. Smith, and J. Pemberton. 1999. Parasite mediated selection against inbred soay sheep in a free-living, island population. Evolution 53:1259-1267.

Coop, R. L., and P. H. Holmes. 1996. Nutrition and parasite interaction. International Journal for Parasitology 26:951-962.

Dalmagro, A. D., and E. M. Vieira. 2005. Patterns of habitat utilization of small rodents in an area of Araucaria forest in Southern Brazil. Austral Ecology 30:353-362.

Decker, K. H., D. W. Duszynski, and M. J. Patrick. 2001. Biotic and abiotic effects on endoparasites infecting *Dipodomys* and *Perognathus* species. Journal of Parasitology 87:300-307.

Delahay, R. J., M. J. Daniels, D. W. Macdonald, K. McGuire, and D. Balharry. 1998. Do patterns of helminth parasitism differ between groups of wild-living cats in Scotland? Journal of Zoology 245:175-183.

Díaz, M., and C. L. Alonso. 2003. Wood mouse *Apodemus sylvaticus* winter food supply: density, condition, breeding, and parasites. Ecology 84:2680-2691.

Díaz, M., T. Santos, and J. L. Tellería. 1999. Effects of forest fragmentation on the winter body condition and population parameters of an habitat generalist, the wood mouse *Apodemus sylvaticus*: a test of hypotheses. Acta Oecologica 20:39-49.

Ehman, K. D., and M. E. Scott. 2002. Female mice mate preferentially with non-parasitized males. Parasitology 125:461-466.

Emmons, L. H., and F. Feer. 1997. Neotropical rainforest mammals: a field guide. 2nd edition. The University of Chicago Press, Chicago.

Fonseca, G. A. B., G. Herrmann, Y. L. R. Leite, R. A. Mittermeier, A. B. Rylands, and J. L. Patton. 1996. Lista anotada dos mamíferos do Brasil. Occasional Papers in Conservation Biology 4:1-38.

Fonseca, G. A. B., and M. C. M. Kierulff. 1989. Biology and natural history of Brazilian Atlantic forest small mammals. Bulletin of the Florida State Museum, Biological Sciences 34:99-152.

Forget, P.-M. 1991. Evidence for secondary seed dispersal by rodents in Panama. Oecologia 87:596-599.

Forget, P.-M. 1993. Post-dispersal predation and scatterhoarding of *Dipteryx panamensis* (Papilionaceae) seeds by rodents in Panama. Oecologia 94:255-261.

Forget, P.-M., T. Milleron, F. Feer, O. Henry, and G. Dubost. 2000. Effects of dispersal pattern and mammalian herbivores on seedling recruitment for *Virola mechelii* (Myristicaceae) in French Guiana. Biotropica 32:452-462.

Galindo-Leal, C., and I. de Gusmão Câmara. 2003. The Atlantic forest of South America: Biodiversity status, threats, and outlook. Island Press, Washington, Covelo, London.

Gillespie, T. R., and C. A. Chapman. 2006. Prediction of parasite infection dynamics in primate metapopulations based on attributes of forest fragmentation. Conservation Biology.

Gillespie, T. R., C. A. Chapman, and E. C. Greiner. 2005. Effects of logging on gastrointestinal parasite infections and infection risk in African primates. Journal of Applied Ecology 42:699-707.

Godoy Teixeira, A. M. d. 2005. Análise da dinâmica da paisagem e de processos de fragmentação e regeneração na região de Caucaia-do-Alto, SP (1962-2000). University of São Paulo, São Paulo.

Gordon, H. M., and H. V. Whitlock. 1939. A new technique for counting nematode eggs in sheep faeces. Journal of the Council for Science and Industrial Research Australia 12:50-52.

Gregory, R. D., and T. M. Blackburn. 1991. Parasite prevavalence and host sample size. Parasitology Today 7:316-318.

Guyer, C. 1988. Food supplementation in a tropical mainland anole, *Norops humilis*: demographic effects. Ecology 69:350-361.

Harf, R., and S. Sommer. 2005. Association between major histocompatibility complex class II DRB alleles and parasite load in the hairy-footed gerbil, *Gerbillus paeba*, in the southern Kalahari. Molecular Ecology 14:85-91.

Haukisalmi, V., and H. Henttonen. 2000. Variability of helminth assemblages and populations in the bank vole *Clethrionomys glareolus*. Polish Journal of Ecology 48:219-231.

Hugot, J.-P., P. Baujard, and S. Morand. 2001. Biodiversity in helminths and nematodes as a field of study: an overview. Nematology 3:199-208.

Janos, D. P., and C. T. Sahley. 1995. Rodent dispersal of vesicular-arbuscular mycorrhizal fungi in Amazonian Peru. Ecology 76:1852-1858.

Kavaliers, M., E. Choleris, and D. W. Pfaff. 2005. Genes, odours and the recognition of parasitized individuals by rodents. TRENDS in Parasitology 21:423-429.

Krebs, C. J. 1966. Demographic changes in fluctuating populations of *Microtus californicus*. Ecological Monographs 36:239-273.

Kristan, D. M., and K. A. Hammond. 2000. Combined effects of cold exposure and sub-lethal intestinal parasites on host morphology and physiology. The Journal of Experimental Biology 203:3495-3504.

Kristan, D. M., and K. A. Hammond. 2001. Parasite infection and caloric restriction induce physiological and morphological plasticity. American Journal of Physiology - Regulatory, Integrative and Comparative Physiology 281:502-510.

Lin, Y.-T. K., and G. O. Batzli. 2001. The influence of habitat quality on dispersal, demography, and population dynamics of voles. Ecological Monographs 71:245-275.

Macdonald, D. W. 1996. Dangerous liaisons and disease. Nature 379:400-401.

Martins, E. G., and V. Bonato. 2004. On the diet of Gracilinanus microtarsus (Marsupialia, Didelphidae) in an Atlantic rainforest fragment in southeastern brazil. Mammalian Biology 69:58-60.

May, R. M., and D. R. Anderson. 1979. Population biology of infectious deseases: Part II. Nature 280:455-461.

Meyer-Lucht, Y., and S. Sommer. 2005. MHC diversity and the association to nematode parasitism in the yellow-necked mouse (Apodemus flavicollis). Molecular Ecology 14:2233-.

Morand, S., and R. Poulin. 1998. Density, body mass and parasite species richness of terrestrial mammals. Evolutionary Ecology 12:717-727.

Murray, D. L., J. R. Cary, and L. B. Keith. 1997. Interactive effects of sublethal nematodes and nutritional status on snowshoe hare vulnerability to predation. Journal of Animal Ecology 66:250-264.

Murray, D. L., L. B. Keith, and J. R. Cary. 1998. Do parasitism and nutritional status interact to affect production in snowshoe hares? Ecology 79:1209-1222.

Musser, G. G., and M. D. Carleton. 1993. Family Muridae. Pages 501-753 in

D. E. Wilson and D. M. Reeder, editors. Mammal species of the world: a taxonomic and geographic reference. Smithsonian Institution Press, Washington and London.

Myers, N., R. A. Mittermeier, C. G. Mittermeier, G. A. B. Fonseca, and J. Kent. 2000. Biodiversity hotspots for conservation priorities. Nature 403:853-858.

Nitikman, L. Z., and M. A. Mares. 1987. Ecology of small mammals in a gallery forest of Central Brazil. Annals of Carnegie Museum 56:75-95.

Nunn, C. L., S. Altizer, K. E. Jones, and W. Sechrest. 2003. Comparative tests of parasite species richness in primates. The American Naturalist 162:597-614.

Oliveira-Filho, A. T., and M. A. L. Fontes. 2000. Patterns of floristic differentiation among Atlantic Forests in Southeastern Brazil and the influence of climate. Biotropica 32:793-810.

Otis, D. L., K. P. Burnham, G. C. White, and D. R. Anderson. 1978. Statistical inference from capture data on closed animal populations. Wildlife Monographs 62:5-135.

Pardini, R., S. Marques de Souza, R. Braga-Neto, and J. P. Metzger. 2005. The role of forest structure, fragment size and corridors in maintaining small mammal abundance and diversity in an Atlantic forest landscape. Biological Conservation 124:253-266.

Paterson, S., W. K., and J. Pemberton. 1998. Major histocompatibility complex variation associated with juvenile survival and parasite resistance in a large unmanaged ungulate population (Ovis aries L.). Proceedings of the National Academy of Sciences of the United States of America 95:3714-3719.

Pimentel, D. S., and M. Tabarelli. 2004. Seed dispersal of the palm Attalea oleifera in a remnant of the

Brazilian Atlantic forest. Biotropica 36:74-84.

Pizo, M. A. 1997. Seed dispersal and predation in two populations of *Cabralea canjerana* (Meliaceae) in the Atlantic forest of southeastern Brazil. Journal of Tropical Ecology 13:559-578.

Poulin, R. 1996. Sexual inequalities in Helminth infections: a cost of being a male? The American Midland Naturalist 147:287-295.

Poulin, R. 1999. The functional importance of parasites in animal communities: many roles at many levels? International Journal for Parasitology 29:903-914.

Ross, J. L. S., and I. C. Moroz. 1997. Mapa Geomorfológico do Estado de São Paulo: escala 1:500.000. FFLCH-USP. IPT and Fapesp, São Paulo.

Rossin, A., and A. I. Malizia. 2002. Relationship between helminth parasites and demographic attributes of a population of the subterranean rodent *Ctenomys talarum* (Rodentia: Octodontidae). Journal of Parasitology 88:1268-1270.

Sánchez-Cordero, V., and R. Martínez-Gallardo. 1998. Postdispersal fruit and seed removal by forest-dwelling rodents in a lowland rainforest in Mexico. Journal of Tropical Ecology 14:139-151.

Schulte-Hostedde, A. I., B. Zinner, J. S. Millar, and G. J. Hickling. 2005. Restitution of mass-size residuals: validating body condition indices. Ecology 86:155-163.

Schwaiger, F. W., D. Gostomski, and M. J. Stear. 1995. An ovine major histocompatibility complex DRB1 allele is associated with low fecal counts following natural, predominantly *Ostertagia circumcincta* infection. International Journal for Parasitology 25:815-822.

Stear, M. J., K. Bairden, J. L. Duncan, P. H. Holmes, Q. A. McKellar, M. Park, S. Strain, M. Murray, S. C. Bishop, and G. Gettinby. 1997. How hosts control worms. Nature 389:27.

Stien, A., R. J. Irvine, E. Ropstad, O. Halvorsen, R. Langvatn, and S. D. Albon. 2002. The impact of gastrointestinal nematodes on wild reindeer: experimental and cross-sectional studies. Journal of Animal Ecology 71:937-945.

Tabarelli, M., L. P. Pinto, J. M. C. Silva, M. Hirota, and L. Bedê. 2005. Challenges and Opportunities for Biodiversity Conservation in the Brazilian Atlantic Forest. Conservation Biology 19:695-700.

Terborgh, J. 1992. Maintenance of diversity in tropical forests. Biotropica 24:283-292.

Tripet, F., and H. Richner. 1997. Host responses to ectoparasites: food compensation by parent blue tits. Oikos 78:557-561.

Tyndale-Biscoe, H. 2005. Life of Marsupials. CSIRO Publishing, Collingwood.

Umetsu, F., and R. Pardini. *in press*. Small mammals in a mosaic of forest remnants and anthropogenic habitats - evaluating matrix quality in an Atlantic forest landscape. Landscape Ecology.

Vieira, E. M., and P. Izar. 1999. Interactions between aroids and arboreal mammals in the Brazilian Atlantic rainforest. Plant Ecology 145:75-82.

Vieira, E. M., M. A. Pizo, and P. Izar. 2003. Fruit and seed exploitation by small rodents of the Brazilian Atlantic forest. Mammalia 67:533-539.

Wilson, K. R., and D. R. Anderson. 1985. Evaluation of two density estimators of small mammal population size. Journal of Mammalogy 66:13-21.

Yuill, T. M. 1987. Diseases as components of mammalian ecosystmes: mayhem and subtlety. Canadian Journal of Zoology 65:1061-1066.

Possible sources of error

THE USE OF SHERMAN TRAPS

Some possible error sources during this study have to be considered. As with any trapping method, capture probabilities often vary for different species or even different individuals of the same species (White et al. 1982). These variations may be due to obvious reasons (e. g. sex, age) or unknown sources of variation (e. g. social dominance, number of traps within the home range of an animal). Further, there may be behavioural reasons for variation such as trap-happiness (i. e. subsequent capture of the same individual in one trap) or trap-shyness (i. e. one single trapping occasion which scares the individual away forever) and time dependence (e. g. increased/ decreased probability due to weather conditions). These effects - which are inherent to any study using traps and bait to locate animals - may lead to shortcomings in the representativeness of the sample. A particular disadvantage of Sherman traps is that capturing one individual prevents others from being captured in the same trap. In other words, if one individual enters the trap early at night, others, which probably would have been caught later that night, will not be recorded. The alternative of using another type of trap, e. g. pitfall traps, was not feasible in our situation. For example, the use of pitfall traps does not allow for collection of faeces samples, because a correct assignment to the animal is impossible in traps containing several individuals. Further, the recapture rate in pitfall traps is very low (Pardini, pers. comm.) and therefore any calculation of distances moved would not have been possible. Therefore, with regard to the hypotheses in focus it was meaningful for us to accept the disadvantages of this type of trapping due to the advantages provided by the use of Sherman traps.

INDEPENDENCE OF INDIVIDUAL CAPTURES

In the investigation of responses of small mammals to vegetation structures (Chapter 3) a possible concern regards the assumption of independence between samples of individual small mammals. When capturing small mammals, independence was assured by only using the first captures of an individual in the analysis. Thereby, possible trap happiness of an individual was not taken into account. Further, the results may be confounded by the fact that possibly one individual was captured in a certain trap location not because of the preferred vegetation structure, but because of the presence (or absence) of another individual in this location. I avoided this problem by not including the number of different individuals but only if at least one individual was captured in a specific location or not. Anyway, neighbouring trap locations may suffer from the same problem. A trap location might be used (or not be used) because of a capture in a neighbouring location. This might have a confounding effect on the experiment and therefore influence the results. To avoid the problem I could either have chosen larger distances between trap locations (which was not feasible due to the investigation of movement distances) or only include some trap locations in the analysis. This would have made the sample size very small due to the fact that for some species not many individuals were captured in some fragments. Therefore, I kept all trap locations in the analysis and accepted the resulting deficiencies.

REFERENCES

White, G. C., D. R. Anderson, K. P. Burnham, and D. L. Otis. 1982. Capture-recapture and removal methods for sampling closed populations. Los Alamos, Los Alamos National Laboratory.

GENERAL DISCUSSION

This study was designed to identify consequences of habitat fragmentation which had an impact on the behaviour or the ecology of focus small mammal species in order to use this information for successful conservation planning.

A major aspect was to identify the differences in response to habitat fragmentation in generalist and specialist species in order to identify the components which mostly influence the most vulnerable species. The investigation concentrated on four main themes, namely the movement distances, density-area-relationship, response to vegetation structure and the differences in parasite load of the different species. So far, knowledge on basic ecological parameters of these selected rodent and marsupial species was very scarce (Fonseca and Kierulff 1989, Barros-Battesti et al. 2000).

The fragmented forest area and agricultural landscape of Caucaia do Alto, Sao Paulo, represented an appropriate study area to investigate influences of habitat fragmentation on small mammals and gain basic ecological information on small mammal species of the Atlantic forest. Small mammals were studied in fragments of different sizes and distances to other fragments but very similar in vegetation, age, climatic conditions and structural appearance of the forest. Thus,

it could be assured that differences in species ecology between fragments were most likely due to the consequences of fragmentation and not to other possible environmental differences.

The use of trapping grids in the different forest fragments made it possible to collect qualitatively very different data. By examining distances between traps where an individual was captured in subsequent nights, I could examine movement behaviour of the focus species. Further, using trapping grids which covered a certain area (in this case 3.2 ha plus boundary strip, see chapter 2), I was able to calculate density estimates for the species. Each point of capture could be characterized in terms of vegetation structural conditions and thus species occurrence linked to these data. This allowed me to identify preferences in vegetation structure of the species. Finally, by using a closed trap type (Sherman traps), I was able to collect individual faeces samples and investigate nematode parasite burden in these rodents and marsupials, which have, to my knowledge, not been subject to an investigation before. Thanks to the provision of the landscape variable data by Jean-Paul Metzger and his lab, Laboratório de Ecologia de Paisagens e Conservação – LEPaC, Departamento de Ecologia, Instituto de Biociências, University of São Paulo, I could link the data to the environmental differences between the investigated forest fragments.

The results described in chapters 1 to 4 exhibited differences between the responses of the investigated species to fragmentation aspects. Here I discuss the results as a whole with respect to the question of the influence of fragmentation.

MOVEMENT DISTANCES, DENSITY AND RESPONSE TO VEGETATION STRUCTURE

Rodents

A. montensis and O. nigripes revealed very similar results in this study. Both are considered generalist species (e. g. Dalmagro and Vieira 2005) and in fact the results underline the hypothesis that generalist species do not show any or only a small response to habitat fragmentation. Both A. montensis and O. nigripes were captured in several different habitats before (Mares et al. 1981, Alho 1982, Alho et al. 1986, Nitikman and Mares 1987, Mares et al. 1989, Stallings 1989, Gentile and Fernandez 1999, Vieira et al. 2004, Dalmagro and Vieira 2005), O. nigripes even in grassland between Atlantic forest fragments (Feliciano et al. 2002). Therefore, I expected this species to show low or none response in behaviour (movements) or density to fragmentation effects. The recorded distances moved fell within the range of other studies for both species (Davis 1945, Nitikman and Mares 1987, Gentile and Cerqueira 1995, Pires et al. 2002). A. montensis showed the lowest distances moved between successive captures and additionally the lowest mean maximum distance moved compared to all investigated species. O. nigripes also had comparably low mean maximum distances moved, especially for females. This might indicate that there is no need for these species to travel long distances for foraging, mating or other possible reasons. This is also shown in the distribution of movement distances (Fig 1, Chapter 1). More than 80 % of their movements did not exceed 50 m. No influence of fragment size on the distance moved was detected. Although this fits in well in the hypothesis of a generalist species, it is to be taken with caution, because none of the investigated species showed any difference in distances moved between fragments of different sizes.

A. montensis differed in mean population density between the fragments but did not show any correlation between fragmentation variables and population density. Density of O. nigripes was not correlated to any fragmentation variables, either. This underlines the hypothesis of generalist species not being effected by fragmentation.

A. montensis seems to be capable of benefiting from the situation of a fragmented landscape. O. nigripes was also caught in non-forest habitat in other studies (e.g. Feliciano et al. 2002) and is not restricted to forest habitat. Therefore its ability to use initial vegetation is not surprising. Both species were captured more frequently in locations characterized by a low canopy and dense understory vegetation (Chapter 3). These results show that these species prefer initial vegetation which is more common in forest edges or possibly within the between-fragment matrix. By increasing the percentage of initial vegetation cover through fragmentation, *A. montensis* and *O. nigripes* might not only benefit from the increase in suitable habitat but also from the absence of other probably competitive species which may not be able to deal with such a change in habitat structure as well as generalist species (Schweiger et al. 1999). Furthermore, both species were able to select the preferred vegetation structure most efficiently and selection was more pronounced in smaller fragments. This underlines the ability of these species to benefit from fragmentation in terms of optimal use of preferred vegetation structures.

In contrast to the two generalist species *A. montensis* and *O. nigripes*, *D. sublineatus* is described as a forest species (Fonseca et al. 1996). Pardini et al. (2005) revealed that *D. sublineatus* was significantly less abundant in medium and small isolated fragments in the study area. Reasons for this vulnerability to habitat change were investigated here. The hypothesis that *D. sublineatus* as a habitat specialist species is more effected by fragmentation could be proved. However, with regard to the mean distances moved no difference between fragments could be detected. On the other hand, *D. sublineatus* showed a positive area-mean density relationship. Further, its mean population density was negatively correlated to the proportion of forest edge in relation to area. This shows clearly that the population of *D. sublineatus* is negatively effected by fragmentation effects, namely habitat loss and edge effects. The investigation of the response to vegetation structure also revealed specialist behaviour of *D. sublineatus*. It was captured more frequently in areas of fragments with a closed canopy, a characteristic of older, less disturbed forest parts (Chapter 3). The results indicate that this species might suffer from the absence of older forest parts from the fragments and the accompanied loss of optimal habitat.

Studies on the density-area relationship of very different taxa revealed inconsistent results (reviews in Bowers and Matter 1997, Connor et al. 2000, Debinski and Holt 2000) with some showing positive, no or negative patch-area-density relationships. Investigations on neotropical primates and small mammals in different ecosystems found no significant relationship between patch-size and density in the majority of species

(Bowers and Matter 1997, Connor et al. 2000). One hypothesis, which was investigated in this study, is that density of specialist species will decrease with fragmentation. This decrease might be caused by several reasons. For example, fragmentation of the habitat might cause disruption of metapopulations, increased demographic and environmental stochasticity as well as inbreeding depression in isolated populations (Lefkovitch and Fahrig 1985, Debinski and Holt 2000, Frankham et al. 2002). In his review, Connor et al. (2000, see also Bowers and Matter 1997) suggested a qualitative difference in density-area relationships between rare and common mammal species with common species having less positive and more negative relations compared to rare species. The results of this study clearly support this hypothesis. The specialist species *D. sublineatus* decreased in density in the smaller fragments, while the density of both generalists *A. montensis* and *O. nigripes* remained uneffected in relation to area.

There are several possible reasons for these findings. Immigration is considered to be an important factor for the population densities of species that do not show strong edge effects (Bowman et al. 2002). Populations in habitat patches that are supported by immigration from populations in other (larger) fragments and therefore are not isolated have higher survival

probabilities compared to isolated populations (Lefkovitch and Fahrig 1985). Results suggest that both *A. montensis* and *O. nigripes* are not strongly effected by edge effects. Immigration and also abundance in the between-fragment-matrix was not investigated here, but one could speculate that the ability of using initial vegetation and the fact that both species were captured in grassland vegetation in other studies (e.g. Feliciano et al. 2002) might increase the ability of these species to immigrate to smaller fragments from bigger forest remnants. This would improve the capability to colonize small fragments and support populations there. The ability to utilize the matrix clearly is an advantage for species in a fragmented landscape (Gascon et al. 1999). In case of those generalist species, the landscape cannot be seen as several habitat patches divided by a hostile environment but rather as a continuum of habitats with different suitability (Andrén 1994). Patchiness of a habitat can only be defined with regard to the requirements of the focal species (Bowers and Matter 1997). This might be the case for the two generalist species *A. montensis* and *O. nigripes* in the study area. The fact, that densities do not differ for these species suggest that performance (measured as density) is not altered by fragmentation. Nevertheless, capture data from the matrix in the study area is needed to prove such an assumption.

On the other hand, other studies showed that the ability of focal species to utilize the matrix and the fragments led to negative density-area relationships (Debinski and Holt 2000). This could not be approved here. The relationship in this study was neither positive nor negative for *A. montensis* and *O. nigripes*. Although there was no apparent relation to area, a significant difference in mean density between fragments was however recorded for *A. montensis*. Highest densities were reached in medium sized forest remnants. This suggests that other factors not recorded here had an influence on the density of this species. Possible other factors might include the presence/absence of predators or the availability of food in certain fragments (Chiarello 2000).

According to the density-area relationship of *D. sublineatus*, higher population density on large habitat patches can be caused by higher reproductive recruitment in these patches (Connor et al. 2000). I did not investigate recruitments rates of *D. sublineatus*, but it might be possible that this species does better in terms of survival and fitness due to environmental circumstances in large fragments and therefore has increased reproductive success. This may lead to higher density in larger fragments. Further, *D. sublineatus* may be less likely to disperse from large habitat patches, resulting in higher densities in larger

patches (Foster and Gaines 1991). Again, dispersal data of *D. sublineatus* is needed to prove this assumption but it is likely that the dispersal ability of this species is low compared to other investigated species.

Because of low capture rates the two other investigated species *O. russatus* and *T. nigrita* were only included in the analysis of movement distances. The results do not allow for conclusions on their vulnerability to fragmentation. Nevertheless, the fact that *O. russatus* was only captured in the biggest fragment and the Morro Grande Reserve and *T. nigrita* only in the Reserve indicates that both species suffer from habitat fragmentation and habitat loss. In other studies, both species were considered as specialist forest species (Olmos 1991).

Marsupials

The two investigated marsupial species differed in their response to fragmentation effects. Analysis of movement distances indicated that both species moved much longer distances in higher proportions of all movements compared to the investigated rodent species. There was no difference between the two species according to movement distances, suggesting that spatial habitat requirements are similar. Differences in mean distances moved inside fragments of different sizes were also not apparent. Obviously, fragmentation does not influence these species in their movement distances. Thus, movement distances do not seem to be a good predictor for fragmentation effects for these species. However, even the smallest investigated fragment was obviously big enough to support these species in their spatial needs. Further decrease in size of the fragments might have consequences, but was not investigated here and further investigation in smaller fragments would be useful.

M. incanus showed an increased trappability in closed canopy locations, indicating a preference of this species for older forest parts. On the other hand, G. microtarsus was captured mostly in locations with an open canopy, suggesting a higher tolerance to disturbed forest parts. This indicates that M. incanus is more sensitive compared to G. microtarsus in terms of habitat use. Vieira and Monteiro-Filho (2003) found a vertical habitat segregation between G. microtarsus and Marmosops paulensis (which is similar in habitus and ecology to M. incanus) with G. microtarsus using upper strata in contrast to M. paulensis, which used neither ground nor understory or canopy significantly more. M. incanus is described also as a scansorial species (e.g. Stallings 1989, Passamani 1995). Laurance (1995) concluded that arboreality might diminish the chance of Australian marsupials to colonize habitat fragments in a patchy landscape. One might assume that the much more

arboreal G. microtarsus compared to M. incanus is more vulnerable to fragmentation. This seems not to be the case in our study because the more arboreal species G. microtarsus did not show a clear negative response to fragmentation effects. On the contary, G. microtarsus seemed to be less effected compared to the scansorial M. incanus. A similar result was achieved in other studies on small mammals in Atlantic forest fragments. Viveiros de Castro and Fernandez (2004) found even a positive correlation between degree of arboreality and number of persisting marsupial populations in forest fragments. However, in my study, M. incanus was captured in all forest fragments suggesting the degree of isolation of forest remnants probably does not inhibit this species from colonizing even isolated, small fragments and the influence of fragmentation effects is not yet critical.

The investigation of population densities in association with fragmentation variables (Chapter 2) revealed no clear relationships between the density of the marsupials and any of the fragmentation variables. Surprisingly, the only association found was between density of M. incanus and the connectivity of the fragments with lower mean density in more connected fragments. This result is contrary to other studies, especially to Pardini et al. (2005) in the same study area and remains obscure. Pardini et al.

(2005) showed that *M. incanus* was more common in native vegetation and connected fragments in the same study area as this study. In another study (Bonvicino et al. 2002) both *G. microtarsus* and *M incanus* were classified as "common but not abundant species" compared to other species captured in the Atlantic forest Biome during the study. In comparison with my results, *M. incanus* was only captured inside secondary forest while *G. microtarsus* was only trapped in "very disturbed" vegetation (Bonvicino et al. 2002). Stallings describes *M. incanus* as inhabiting both primary and secondary forest habitat (Stallings 1989) and captured *G. microtarsus* in intermediately disturbed habitat. Taking these and my results into account, it seems that *G. microtarsus* is less specialized in relation to habitat requirements compared to *M. incanus*. However, both species have not been captured in matrix habitat (Feliciano et al. 2002, Viveiros de Castro and Fernandez 2004) which is an important criterion for long time persistence in fragmented habitats, indicating that probably further isolation of fragments might have stronger effects on these species.

PARASITE BURDEN

The investigation of the parasite load and body condition of the focus species

and possible influences of fragmentation on these ecologically important factors revealed unexpected results. The hypothesis that specialists decrease in body condition with increasing fragmentation and increase in parasite load in contrast to unimpaired generalists could not be proven. In fact, the specialist rodent *D. sublineatus* was the only species for which a comprehensible association of condition, parasite load, and landscape variables could be detected. But against expectations, the main factor that seemed to influence condition index and parasite load was not the degree of fragmentation but population density of this species. Decreased fragmentation effects in larger fragments induced higher population density for this species (Chapter 2) which was correlated negatively to body condition and positively to parasite load. It is likely that enhanced density improved conditions for transmission of directly transmitted parasites and thereby caused this pattern. It has been shown that host density increases parasite prevalence and diversity (May and Anderson 1979, Morand and Poulin 1998, Gillespie et al. 2005, Kavaliers et al. 2005). Relatively low body condition might even have improved circumstances for parasites by reducing the ability of immune response (May and Anderson 1979, Poulin 1996).

In *M. incanus* a similar pattern was observed in relation to the condition index and landscape variables but a contradictory association between density and body condition. The body condition decreased in fragments with lower density. Detection of relationship between parasite load and density failed. Possibly, higher densities improve conditions for this species, which in turn might result in a better overall body condition. This is, for example, easily conceivable in species that hunt in packs and therefore increase the chance of successful hunting, but difficult to explain for a small solitary species feeding mainly on insects. Clearly, further investigations on the body condition of these species are needed.

In *G. microtarsus*, body condition improved with increasing edge density which underlines the ability of this species to benefit from fragmentation. Possible reasons for this finding may be a higher abundance of food items - mainly insects (Martins and Bonato 2004) - in fragments with higher edge density, although this would have to be tested in an appropriately designed experiment.

However, the hypothesis that generalist species are less influenced by fragmentation in parasite load and body condition compared to specialist species could be approved. For the two generalist rodents *A. montensis* and *O. nigripes* no associations between fragmentation parameters and neither parasite burden nor body condition could be detected; except for the higher parasite burden of *O.*

nigripes in fragments with increased edge density. As stated before, if these species are able to cross or even utilize the matrix as habitat, there is no fragmentation of habitat apparent and the whole area can rather be seen as a habitat with different suitability. In this sense, it seems obvious that forest fragmentation does not have any impact on conditional status or parasite burden of these species.

CONCLUSIONS

The results in this study showed the differences between species in relation to the requirements within the same habitat, the coastal Atlantic forest. Within the context of conservation, results underline the importance of considering the special needs of the respective species in focus. Which conservation strategy is most effective depends on the special ecology and the individual requirements of the species in question (Connor et al. 2000). In most cases it is not possible to conserve the entire habitat or areas comprising several habitats but only parts of it (or parts of a habitat). In the case of the remaining Atlantic forest, it is in most areas only possible to conserve certain forest fragments as the whole forest is for the most part composed out of forest fragments of different sizes (Terborgh 1992, Ranta et al. 1997, Galindo-Leal and de Gusmão Câmara 2003, Tabarelli et al. 2005). Additionally, the majority of fragments consist of secondary forest, including the study area in this investigation (Godoy Teixeira 2005). Due to limited (financial) resources for conservation, decisions must be made about which fragment to conserve in order to preserve a maximum diversity, or more precisely to preserve the species, which are dependent on the forest habitat and are most threatened by extinction in case of further destruction of the Atlantic forest. The results of this study encourage the protection of "single large" forest remnants instead of "several small" (the so-called "SLOSS" debate; Diamond et al. 1976, Simberloff and Abele 1976, Wilcox 1980, Wilcox and Murphy 1985, Saunders et al. 1991, Bierregaard et al. 1992, Tscharntke et al. 2002). The investigations in the study area of Caucaia presented that for specialized forest species like *D. sublineatus* and, to a somewhat smaller extent, *M. incanus,* survival is much more likely in habitats provided by large forest remnants, which includes an appropriate structure of the forest. This is demonstrated by the decline in population density of the specialist *D. sublineatus*. In contrast, for the generalist species *A. montensis* and *O. nigripes,* such considerations are not needed because of their ability to survive even in small and very disturbed fragments without any detected disadvantages. Studies on fragmentation effects revealed similar

results for endemic species. Fragmentation led to decrease in abundance or survival of endemics in different forest ecosystems (review in Turner 1996). For example, in another biodiversity hotspot, the tropical rainforest of the Western Ghats in southern India (Myers et al. 2000), human disturbance and fragmentation of original rainforest habitat led, among other consequences, to loss of specialized endemic small mammals and reptiles (Kumar et al. 2002). The same critical pattern was observed for different taxa of animals, including birds (e.g Newmark 1991), amphibians (e.g. McDonald and Alford 1999, Green 2003), insects (e.g. Gibbs and Stanton 2001, Baguette and Schtickzelle 2003) and plant species (Martinez-Garza and Howe 2003 and references therein).

If the goal in conservation is to maintain populations of specialized and comparably rare species like *D. sublineatus* that are at least larger than some minimum viable population size, results suggest that this goal is a lot easier to achieve through the conservation of large forest remnants than several small ones. Smaller remnants are much more prone to disturbances and influences from outside the fragment. Larger fragments have automatically a greater core area which is free or at least much less influenced from edge effects (Saunders et al. 1991). Therefore, species suffering from edge effects are less prone to such disturbances in larger fragments. Further, species which are not able to maintain a metapopulation structure through the ability to cross the matrix and (re)colonize more isolated fragments need larger forest remnants to maintain a population large enough to be protected from demographic stochiasticity effects, environmental effects and loss of genetic variability (e.g. Burkey 1995, Frankham et al. 2002).

ONGOING STUDIES

This study was accomplished in a landscape comprising an amount of forest habitat of approximately 31 % embedded in non-forest (mostly agricultural) land. Recent studies suggest that close to this amount of original habitat a threshold exists, below which the impact of fragmentation and habitat isolation as well as the importance of habitat spatial configuration for species occurrence increases drastically (Andrén 1994, With and Crist 1995, Fahrig 1997, Fahrig 2003). Based on the knowledge and the experiences of our Brazilian co-operators, we intend to investigate in ongoing studies the influence of different proportions of forest habitat in a landscape on small mammal population dynamics. Therefore we plan to capture small mammals in different regions of the coastal Atlantic forest comprising different amounts (15 %, 30 %, 45 % and 100 %) of forest cover. By

this, we intend to examine the relative importance of the amount of forest cover for the population dynamics of small mammal populations and endangered species of the Atlantic forest.

REFERENCES

Alho, C. J. R. 1982. Brazilian rodents: their habitats and habits. Mammalian biology in South America 6:143-166.

Alho, C. J. R., L. A. Pereira, and A. C. Paula. 1986. Patterns of habitat utilization by small mammal populations in cerrado biome of central Brazil. Mammalia 50:447-460.

Andrén, H. 1994. Effects of habitat fragmentation on birds and mammals in landscapes with different proportions of suitable habitat: a review. Oikos 71:355-366.

Baguette, M., and N. Schtickzelle. 2003. Local population dynamics are important to the conservation of metapopulations in highly fragmented landscapes. Journal of Applied Ecology 40:404-412.

Barros-Battesti, D. M., R. Martins, C. R. Bertim, N. H. Yoshinari, V. L. N. Bonoldi, E. P. Leon, M. Miretzki, and T. T. S. Schumaker. 2000. Land fauna composition of small mammals of a fragment of Atlantic forest in the state of Sao Paulo, Brazil. Revista Brasileira de Zoociencias 17:241-249.

Bierregaard, R. O., T. E. Lovejoy, K. Valerie, A. A. dos Santos, and R. W. Hutchings. 1992. The biological dynamics of tropical Rainforest fragments: a prospective comparison of fragments and continuous forest. BioScience 42:859-866.

Bonvicino, C. R., S. M. Lindbergh, and L. S. Maroja. 2002. Small non-flying mammals from conserved and altered areas of Atlantic Forest and Cerrado: comments on their potential use for monitoring environment. Brazilian Journal of Biology 62:765-774.

Bowers, M. A., and S. F. Matter. 1997. Landscape Ecology of mammals: relationships between density and patch size. Journal of Mammalogy 78:999-1013.

Bowman, J., N. Cappuccino, and L. Fahrig. 2002. Patch Size and Population Density: the Effect of Immigration Behavior. Conservation Ecology 6: http://www.consecol.org/vol6/iss1/art9.

Burkey, T. V. 1995. Extinction rates in archipelagoes: Implications for populations in fragmented habitats. Conservation Biology 9:527-541.

Chiarello, A. G. 2000. Density and population size of mammals in remnants of brazilian Atlantic forest. Conservation Biology 14:1649-1657.

Connor, E. F., A. C. Courtney, and J. M. Yoder. 2000. Individuals-area relationships: the relationship between animal population density and area. Ecology 81:734-748.

Dalmagro, A. D., and E. M. Vieira. 2005. Patterns of habitat utilization of small rodents in an area of Araucaria forest in Southern Brazil. Austral Ecology 30:353-362.

Davis, D. E. 1945. Home ranges of some brazilian mammals. Journal of Mammalogy 26:119-127.

Debinski, D. M., and R. D. Holt. 2000. A survey and overview of habitat fragmentation experiments. Conservation Biology 14:342-355.

Diamond, J. M., J. Terborgh, R. F. Whitcomb, J. F. Lynch, P. A. Opler, C. S. Robbins, D. S. Simberloff, and L. G. Abele. 1976. Island biogeography and conservation: strategy and limitations. SCIENCE 193:1027-1032.

Fahrig, L. 1997. Relative effects of habitat loss and fragmentation on population extinction. Journal of Wildlife Management 61:603-610.

Fahrig, L. 2003. Effects of habitat fragmentation on biodiversity. Annual Review of Ecology, Evolution and Systematic 34:487-515.

Feliciano, B. R., F. A. S. Fernandez, D. de Freitas, and M. S. L. Figueiredo. 2002. Population dynamics of small rodents in a grassland between fragments of Atlantic forest in southeastern Brazil. Mammalian Biology 67:304-314.

Fonseca, G. A. B., G. Herrmann, Y. L. R. Leite, R. A. Mittermeier, A. B. Rylands, and J. L. Patton. 1996. Lista anotada dos mamíferos do Brasil. Occasional Papers in Conservation Biology 4:1-38.

Fonseca, G. A. B., and M. C. M. Kierulff. 1989. Biology and natural history of brazilian Atlantic forest small mammals. Bulletin of the Florida State Museum, Biological Sciences 34:99-152.

Foster, J., and M. S. Gaines. 1991. The effects of a successional habitat mosaic on a small mammal community. Ecology 72:1358-1375.

Frankham, R., J. D. Ballou, and D. A. Briscoe. 2002. Introduction to Conservation Genetics. Cambridge University Press, Cambridge.

Galindo-Leal, C., and I. de Gusmão Câmara. 2003. The Atlantic forest of South America: Biodiversity status, threats, and outlook. Island Press, Washington, Covelo, London.

Gascon, C., T. E. Lovejoy, R. O. Bierregaard, J. R. Malcolm, P. C. Stouffer, H. L. Vasconcelos, W. F. Laurance, B. L. Zimmerman, M. Tocher, and S. Borges. 1999. Matrix habitat and species richness in tropical forest remnants. Biological Conservation 91:223-229.

Gentile, R., and R. Cerqueira. 1995. Movement patterns of five species of small mammals in a Brazilian restinga. Journal of Tropical Ecology 11:671-677.

Gentile, R., and F. A. S. Fernandez. 1999. Influence of habitat structure on a streamside small mammal community in a Brazilian rural area. Mammalia 63:29-40.

Gibbs, J. P., and E. J. Stanton. 2001. Habitat fragmentation and arthropod community change: carrion beetles, phoretic mites, and flies Ecological Applications 11:79-85.

Gillespie, T. R., C. A. Chapman, and E. C. Greiner. 2005. Effects of logging on gastrointestinal parasite infections and infection risk in African primates. Journal of Applied Ecology 42:699-707.

Godoy Teixeira, A. M. d. 2005. Análise da dinâmica da paisagem e de processos de fragmentação e regeneração na região de Caucaia-do-Alto, SP (1962-2000). University of São Paulo, São Paulo.

Green, D. M. 2003. The ecology of extinction: population fluctuation and decline in amphibians. Biological Conservation 111:331-343.

Kavaliers, M., E. Choleris, and D. W. Pfaff. 2005. Genes, odours and the recognition of parasitized individuals by rodents. TRENDS in Parasitology 21:423-429.

Kumar, A., R. Chellam, B. C. Choudhury, D. Mudappa, K. Vasudevan, N. M. Ishwar, and B. Noon. 2002. Impact of rainforest fragmentation on the small mammals and herpetofauna in the Western Ghats, South India. Wildlife Institue of India, Dehra Dun.

Laurance, W. F. 1995. Extinction and survival of rainforest mammals in a fragmented tropical landscape. Pages 46-63 in W. Z. Lidicker, editor. Landscape approaches in mammalian Ecology and Conservation. University of Minnesota Press, Minneapolis.

Lefkovitch, L. P., and L. Fahrig. 1985. Spatial characteristics of habitat patches and population survival. Ecological Modelling 30:297-308.

Mares, M. A., J. K. Braun, and D. Gettinger. 1989. Observations on the distribution and ecology of the mammals of the cerrado grassland of central Brazil. Annals of Carnegie Museum 58:1-60.

Mares, M. A., R. A. Ojeda, and M. P. Kosco. 1981. Observations on the distribution and ecology of the mammals of Salta Province, Argentina. Annals of Carnegie Museum 50:151-206.

Martinez-Garza, C., and H. F. Howe. 2003. Restoring tropical diversity: beating the time tax on species loss. Journal of Applied Ecology 40:423-429.

Martins, E. G., and V. Bonato. 2004. On the diet of *Gracilinanus microtarsus* (Marsupialia, Didelphidae) in an Atlantic rainforest fragment in southeastern brazil. Mammalian Biology 69:58-60.

May, R. M., and D. R. Anderson. 1979. Population biology of infectious deseases: Part II. Nature 280:455-461.

McDonald, K., and R. Alford. 1999. A review of declining frogs in Northern Queensland. Pages 14-22 in A. Campbell, editor. Declines and disappearances of Australian frogs. Environment Australia, Canberra.

Morand, S., and R. Poulin. 1998. Density, body mass and parasite species richness of terrestrial mammals. Evolutionary Ecology 12:717-727.

Myers, N., R. A. Mittermeier, C. G. Mittermeier, G. A. B. Fonseca, and J. Kent. 2000. Biodiversity hotspots for conservation priorities. Nature 403:853-858.

Newmark, W. D. 1991. Tropical forest fragmentation and the local extinction of understory birds in the eastern Usambara Mountains, Tanzania. Conservation Biology 5:67-78.

Nitikman, L. Z., and M. A. Mares. 1987. Ecology of small mammals in a gallery forest of Central Brazil.

Annals of Carnegie Museum 56:75-95.

Olmos, F. 1991. Observations on the behaviour and population dynamics of some brazilian Atlantic forest rodents. Mammalia 55:555-565.

Pardini, R., S. Marques de Souza, R. Braga-Neto, and J. P. Metzger. 2005. The role of forest structure, fragment size and corridors in maintaining small mammal abundance and diversity in an Atlantic forest landscape. Biological Conservation 124:253-266.

Passamani, M. 1995. Vertical stratification of small mammals in Atlantic hill forest. Mammalia 59:276-279.

Pires, A. S., P. Koeler Lira, F. A. S. Fernandez, G. M. Schittini, and L. C. Oliveira. 2002. Frequency of movements of small mammals among Atlantic coastal forest fragments in Brazil. Biological Conservation 108:229-237.

Poulin, R. 1996. Sexual inequalities in Helminth infections: a cost of being a male? The American Midland Naturalist 147:287-295.

Ranta, P., T. Blom, J. Niemelä, E. Joensuu, and M. Siitonen. 1997. The fragmented Atlantic rain forest of Brazil: size, shape and distribution of forest fragments. Biodiversity and Conservation 7:385-403.

Saunders, D. A., R. J. Hobbs, and C. R. Margules. 1991. Biological consequences of ecosystem fragmentation: a review. Conservation Biology 5:18-32.

Schweiger, E. W., J. E. Diffendorfer, R. Pierotti, and R. D. Holt. 1999. The relative importance of small scale and landscape-level heterogeneity in structuring small mammal communities. Page 347 in G. W. Barrett and J. D. Peles, editors. Landscape ecology of small mammals. Springer, New York.

Simberloff, D. S., and L. G. Abele. 1976. Island biogeography theory and practice. SCIENCE 191:285-286.

Stallings, J. R. 1989. Small mammal inventories in an eastern brazilian park. Bulletin of the Florida State Museum, Biological Sciences 34:159-200.

Tabarelli, M., L. P. Pinto, J. M. C. Silva, M. Hirota, and L. Bedê. 2005. Challenges and Opportunities for Biodiversity Conservation in the Brazilian Atlantic Forest. Conservation Biology 19:695-700.

Terborgh, J. 1992. Maintenance of diversity in tropical forests. Biotropica 24:283-292.

Tscharntke, T., I. Steffen-Dewenter, A. Kruess, and C. Thies. 2002. Contribution of small habitat fragments to conservation of insect communities of grassland-cropland landscapes. Ecological Applications 12:354-363.

Turner, I. M. 1996. Species loss in fragments of tropical rain forest: a review of the evidence. Journal of Applied Ecology 33:200-209.

Vieira, E. M., and E. L. A. Monteiro-Filho. 2003. Vertical stratification of small mammals in the Atlantic rain forest of south-eastern Brazil. Journal of Tropical Ecology 19:501-507.

Vieira, M. V., C. E. V. Grelle, and R. Gentile. 2004. Differential trappability of small mammals in three habitats of Southeastern Brazil. Brazilian Journal of Biology 64:895-900.

Viveiros de Castro, E. B., and F. A. S. Fernandez. 2004. Determinants of differential extinction vulnerabilities of small mammals in Atlantic forest fragments in Brazil. Biological Conservation 119:73-80.

Wilcox, B. A. 1980. Insular Ecology and Conservation. Pages 95-117 in M. E. Soulé and B. A. Wilcox, editors. Conservation Biology: An Evolutionary-Ecological Perspective. Sinauer Associates, Sunderland, Massachusetts.

Wilcox, B. A., and D. D. Murphy. 1985. Conservation strategy: the effects of fragmentation on extinction. The American Naturalist 125:879-887.

With, K. A., and T. O. Crist. 1995. Critical thresholds in species' response to landscape structure. Ecology 76:2446-2459.

ACKNOWLEDGEMENTS

I thank PD Dr. Simone Sommer for giving me the opportunity to do this work and for her scientific guidance throughout the study. Further, without the help of Dr. Renata Pardini this work would not have been possible and I thank her a lot for scientific assistance, the very good cooperation as well as teaching me Brazilian culture. I thank Prof. Dr. Jean-Paul Metzger for all the help and making landscape data available to me as well as Prof. Dr. Jörg U. Ganzhorn for a lot of statistical help and his support as a referee of the dissertation.

I am indebted to Yvonne Meyer-Lucht for a lot of support in several scientific and real-life questions. Furthermore, I thank Fabiana Umetsu for teaching me patiently how to deal with the small mammals of the Atlantic forest, and Dr. Christoph Knogge for the enormous help whenever whatever problem emerged (starting from "The car is crashed!" until "Where can I buy a guitar?").

Special thanks go out to several colleagues and helpers in the field, which I spend most of the time with and which became friends to me: Henning Steinicke, Camilla Armstrong, Claudia Regina Guimarães, Laura Regina Capelari Naxara, Rafael Pimentel, Marcelo Awade, Daniel Pereira Monari, Danilo Boscolo, Sergio Marques de Souza, Stefanie Rüter, Sigrid Keiser, Alexandre Camargo Martensen, Anne Leppin, Laura Sandberger, Jana Jahnke and Julio Mendivel, Juan Pablo Nieto Holgun, Jürg Brendan Logue, and Daniel Piechowski. I thank Jan Axtner, Celine Otten, Esther Verjahns, and Ann Parplys for the parasitological examinations.

Thanks to the whole family of Claudia and Arcanjo Ribeiro de Lima for serving us very good traditional Brazilian food and cultural enrichment. In this context, I thank also Christian Leidenberger, Camilla Armstrong, and Claudia Regina Guimarães for the unofficial German-Brazilian cooking contest, which gave me and others the pleasure to taste very good plates of both countries.

I thank the whole Department of Animal Ecology and Conservation of the University of Hamburg and especially my roommates Dr. Petra Lahann and Yvonne Meyer-Lucht for the good atmosphere.

I thank Delia for everything.

I am thankful to the Bundesministerium für Bildung und Forschung (BMBF) for funding the Project and my work.

Thanks to all others that helped me in whatever manner and which are not mentioned here.

In memory of Tatiana Raso de Moraes Possato.